CIRIA C520

Sealants
microbiological deterioration under wet conditions

C S Dow
D W Aubrey
S Hurley

CIRIA *sharing knowledge ■ building best practice*

6 Storey's Gate, Westminster, London SW1P 3AU
TELEPHONE 020 7222 8891 FAX 020 7222 1708
EMAIL enquiries@ciria.org.uk
WEBSITE www.ciria.org.uk

Summary

This book focuses upon the microbiological degradation of civil engineering sealants under aqueous challenge conditions.

An evaluation procedure has been devised to assess the susceptibility of a sealant formulations to microbiological degradation.

Various sealant formulations were prepared as strips and joints and challenged with defined mixtures of micro-organisms and selected water industry environments. Assessment of sealant compromise was based upon tensile testing using modulus and extensibility as the performance indices together with scanning electron and light microscopy to evaluate the extent of microbial degradation as shown by modification of surface topography. From these data a substantial reference database has been obtained, which permits comparison of the performance of different sealant formulations against defined microbial consortia and by extrapolation to their performance in particular environments.

Definition of the sealant formulations, the application of standard evaluation protocols and the reference database has laid the foundation for meaningful assessment of sealants and their susceptibility to microbiological degradation.

Sealants – microbiological deterioration under wet conditions

Dow, C S, Aubrey, D W, Hurley, S

Construction Industry Research and Information Association

CIRIA C520   © CIRIA 2001   ISBN 0 86017 520 0

Keywords	
Sealants, sealant joints, water treatment plant, sewage plant, polyurethanes, polysulphides, silicones, deterioration, microbiological attack.	

Reader interest	Classification	
Design, specification, materials, construction supervising and operations engineers involved in the water industry	Availability	Unrestricted
	Contents	Best practice
	Status	Committee-guided
	User	Construction/water industry professionals

Published by CIRIA. All rights reserved. No part of this publication may be reproduced or transmitted in any form or by any means, including photocopying and recording, without the written permission of the copyright-holder, application for which should be addressed to the publisher. Such written permission must also be obtained before any part of this publication is stored in a retrieval system of any nature.

Use of this document

This book presents data on the microbiological degradation of selected polymeric sealants.

An evaluation protocol has been devised that enables comparison of sealant performance, using defined performance indices, with a substantial reference database derived from microbiological challenge of defined polymers under defined conditions in the laboratory. Comparison of these data with those from selected environments facilitates laboratory-based evaluation of the susceptibility of sealant formulations to microbiological degradation.

Since, the number of variables inherent in any one challenge situation (chemical, physical or microbiological) are considerable, the procedure presented has attempted to minimise these variables, so permitting a reproducible and robust assessment of sealant performance.

The nature of the study and the data presented leads inevitably to a complex document. The ultimate value of the study lies in the ability to reference future evaluations to the reported database.

Foreword

CIRIA's research programme, "Sealants – microbiological deterioration under wet conditions", has been undertaken to assess the effects of a range of micro-organisms and aqueous environments on selected sealant formulations. A test procedure based on sealant strips has been devised and compared with the general method for tensile testing of concrete/sealant joints.

This report is the result of research carried out under contract to CIRIA by Microbial Systems Ltd in partnership with D W Aubrey, Polymer Chemist and S Hurley, Taywood Engineering Ltd.

Following CIRIA's usual practice, the research project was guided by a steering group, which comprised:

R A Sykes (chairman)	Wimpey Environmental
K Cockayne	Servicised Limited
K Colquhoun	Thames Water Ltd
B W Gill	TEQUA
S Harwood	Severn Trent Water Limited
M Hubbard	SouthWest Water Services Ltd
C Hunter	Building Research Establishment
O Parry	Yorkshire Water Services Limited
R Peck	Anglia Water Services Limited
C Phillips	Bechtel Water Technology Limited
K Preston	Northumbrian Water Limited
T Rees	Morton International
R Robinson	Morton International
K Seal	Thor Chemicals (UK) Limited
N Tarbet	Water Research Centre
T Wilkins/S Zaidi	FOSROC International Technology

CIRIA's research manager was Dr B W Staynes.

ACKNOWLEDGEMENTS

The project was funded by the sponsors of CIRIA's Core Programme, Anglian Water plc, Northumbrian Water plc, Bechtel Water Technology Limited, Severn Trent Water plc, South West Water plc, Yorkshire Water plc, FOSROC International Limited, Morton International Limited.

CIRIA and the authors gratefully acknowledge the support of these funding organisations and the technical help and advice provided by the members of the steering group and the many individuals who were consulted. Contributions do not imply that individual funders necessarily endorse all views expressed in the published output.

Contents

List of figures 7
List of tables 8
Glossary 10
Abbreviations 12
Notation 13
Bibliography 14

1 INTRODUCTION 15
    1.1 Aims and objectives 15
    1.2 Microbial biofilms 15
    1.3 Microbial degradation of sealants 16

2 SEALANT COMPOSITION AND MICROBIAL DEGRADATION ... 21
    2.1 Sealants in general 21
    2.2 Polyurethane sealants 27
    2.3 Polysulfide sealants 30
    2.4 Silicone sealants 32

3 TEST PROCEDURES 35
    3.1 Test specimen geometry and preparation 35
    3.2 Physical test methods 36
    3.3 Microbiological test methods 37
    3.4 Evaluation of results 38

4 EXPERIMENTAL DATA AND INTERPRETATION 47
    4.1 Reported incidents of microbial degradation 47
    4.2 Results obtained with polyurethane sealant strips 49
    4.3 Results obtained with polyurethane sealant joints 51
    4.4 Results obtained with polysulfide strips 54
    4.5 Results obtained with polysulfide sealant joints 56
    4.6 Results obtained with silicone sealant strips 59
    4.7 Results obtained with silicone sealant strips and joints 61
    4.8 WIS ratings and failure modes observed with sealant joints 62
    4.9 Microbiological interpretation 63

5 GENERAL DISCUSSION 67
    5.1 Introduction 67
    5.2 State of cure of sealant before immersion 67
    5.3 Field and laboratory exposure to micro-organisms 68
    5.4 Effects of some sealant variables 69
    5.5 Suitability of laboratory testing for prediction of service performance 71
    5.6 Establishing suitable sealant formulations for use in specific environments . 74

6		SUMMARY, RECOMMENDATIONS AND FUTURE WORK	75
	6.1	Assessment of the susceptibility of sealant formulations to microbial degradation: test procedure	75
	6.2	Assessment of the susceptibility of sealant formulations to microbial degradation: microbiological challenge	75
	6.3	Assessment of the susceptibility of sealant formulations to microbial degradation: evaluation	76
	6.4	Assessment of the susceptibility of sealant formulations to microbial degradation: sealant formulations	76
	6.5	Assessment of the susceptibility of sealant formulations to microbial degradation: joints	76
	6.6	Future work	77
7		RESEARCH DATA SUMMARIES	79
8		REFERENCES	89

APPENDICES

A1		RECOMMENDED TEST PROCEDURES	93
	A1.1	Sealant strips	93
	A1.2	Joints	96
A2		SEALANT FORMULATIONS	101
	A2.1	Sealant strips	101
	A2.2	Joint formulations	102
A3		MICROBIAL CONSORTIA AND CHALLENGE ENVIRONMENTS	103
	A3.1	Microbial consortia	103
	A3.2	Environmental conditions	104
	A3.3	Correlation between environmental conditions and consortia	105
A4		SUMMARY DATA SHEETS: SEALANT STRIPS	107
A5		SUMMARY DATA SHEETS: JOINTS	117

Figures

Figure 1.1	Biofilm development phases	16
Figure 1.2	Potential microbial interactions on a sealant/concrete joint in an aqueous environment	17
Figure 1.3	Outline strategy for the selection of degradative microbes	19
Figure 2.1	Possible modulus change with curing, immersion and microbial attack for a sealant fully cured before immersion	24
Figure 2.2	Possible modulus change of a moisture-curable sealant when it is immersed before completion of dry-curing	24
Figure 3.1	Microbial challenge of test strips	38
Figure 3.2	Assembly of test strips for environmental challenge	39
Figure 3.3	Relationship between σ_{25} and the exposure time of PU2 to different microbial consortia	40
Figure 3.4	Method of evaluating strip test results for σ_{25}	40
Figure 3.5	Changes in surface morphology of sealant strips: (a) polyurethane 1, wet control, six months; (b) polyurethane 1, dry control, six months; (c) hole formation – polyurethane 1 challenged with consortium 2, three months; (d) hole formation – polyurethane 1 challenged with consortium 2, six months; (e) crack formation leading to crumbling – polyurethane 1, Tewkesbury pre-treatment, three months; (f) crack formation leading to crumbling – polyurethane 1, Tewkesbury pre-treatment, six months; (g) spongy effect – polyurethane 1, 99 per cent humidity/sewage, six months; (h) spongy effect – polyurethane 1, consortium 1 challenged in soil, six months	42
Figure 4.1	Compromised sealant at Sunnyside Storage Reservoir	48
Figure 5.1	Use of six time-points	74
Figure A1.1	Data from Davenport-Nene tensile strength testing instrument	94
Figure A1.2	Relationship between σ_{25} and the exposure time of PU2 to different microbial consortia	95
Figure A1.3	Method of evaluating strip test results for σ_{25}	96
Figure A1.4	Design of test specimen used for all joint tests	97
Figure A1.5	Changes to modulus and extensibility after immersion	99
Figure A5.1	Changes to modulus and extensibility after immersion	117

Tables

Table 4.1	Analysis of sealant from Sunnyside and Diamond Avenue reservoirs compared to a typical lead dioxide cured polysulfide sealant	48
Table 4.2	Changes in (a) 25 per cent modulus ($\Delta\sigma_{25}$) and (b) extension at maximum force (ΔE_{max}) relative to the wet controls, for polyurethane sealant joints exposed to various environments	52
Table 4.3	Comparison of (a) $\Delta\sigma_{25}$ and (b) ΔE_{max} results for PU3 in strip and joint form	53
Table 4.4	Changes in (a) 25 per cent modulus ($\Delta\sigma_{25}$) and (b) maximum extension (ΔE_{max}), relative to the wet controls, for polysulfide sealant joints exposed to various environmental conditions	57
Table 4.5	Comparison of (a) σ_{25} or $\Delta\sigma_{25}$ and (b) ΔE_{max} results for PS1 in strip and joint form	58
Table 4.6	Changes in (a) 25 per cent modulus ($\Delta\sigma_{25}$) and (b) maximum extension (ΔE_{max}), relative to the wet controls, for (a) Si2/acrylate joints and (b) Si2 sealant strips exposed to various environmental conditions	61
Table 4.7	Correlation between consortia (C1 to C11) and environmental challenges for each sealant type	65
Table 7.1	Analysis of 20 per cent error bar charts of σ_{25} data for sealant strips exposed to defined microbial consortia for six months	80
Table 7.2	Analysis of 20 per cent error bar charts of sealant strips exposed to different environments for six months	81
Table 7.3	Polyurethane: analysis of 20 per cent error bar charts of σ_{25} data for sealant strips exposed to defined microbial consortia and environments for six months	82
Table 7.4	Polysulfide and silicone: analysis of 20 per cent error bar charts of σ_{25} data for sealant strips exposed to defined microbial consortia for six months	83
Table 7.5	Summary of the surface morphology of sealant strips exposed to different microbial consortia as ascertained by light and scanning electron microscopy	84
Table 7.6	Summary of the surface morphology of sealant strips exposed to different environments as ascertained by light and scanning electron microscopy	85
Table 7.7	Comparison of σ_{25} values for polyurethane strips after six months with the wet control for microbial consortia and environmental challenges	86
Table 7.8	Comparison of σ_{25} values for polysulfide and silicone strips after six months with the wet control for microbial consortia and environmental challenges	87
Table 7.9	a) σ_{25} data for sealant strips from microbial consortia and environmental challenges after six months' exposure; (b) comparison of σ_{25} data for sealant strips after six months' exposure with wet controls	88

Table A2.1	Sealant formulations: polyurethane sealant strips	101
Table A2.2	Sealant formulations: oxime-cured silicone sealant strips	101
Table A2.3	Sealant formulations: benzamide-cured silicone sealant strips	101
Table A2.4	Sealant formulations: polysulfide sealant strips	102
Table A2.5	Joint formulations	102
Table A3.1	Microbial consortia	103
Table A3.2	Characteristics of raw water to Albert Treatment Works and upland treated water of Stansfield Service Reservoir	104
Table A3.3	Correlation between environmental challenge and specific consortia as judged by the effect on sealant performance and surface topography	105
Table A4.1	Polyurethane – consortia – summary of first time-point mean results (three months' exposure)	108
Table A4.2	Polyurethane – consortia – summary of second time-point mean results (six months' exposure)	109
Table A4.3	Polyurethane – environmental – summary of first time-point mean results (three months' exposure)	110
Table A4.4	Polyurethane – environmental – summary of second time-point mean results (six months' exposure)	111
Table A4.5	Silicone/polysulfide – consortia – summary of first time-point mean results (three months' exposure)	112
Table A4.6	Silicone/polysulfide – consortia – summary of second time-point mean results (six months' exposure)	113
Table A4.7	Silicone/polysulfide – environmental – summary of first time-point mean results (three months' exposure)	114
Table A4.8	Silicone/polysulfide – environmental – summary of second time-point mean results (six months' exposure)	115
Table A5.1	Summary of first time-point mean results (three months' exposure)	118
Table A5.2	Summary of second time-point mean results (six months' exposure)	119
Table A5.3	Change in properties compared to dry controls	120
Table A5.4	Change in properties compared to wet controls	121
Table A5.5	WIS rating against dry controls	122
Table A5.6	WIS rating against wet controls	123

Glossary

Adhesive failure	Failure of a joint at (or very close to) the sealant-primer or primer-substrate interface.
Biodegradation	Deterioration or degradation of materials (natural, synthetic or refined) and structures caused by biological activity, eg micro-organisms. The effects may include corrosion, fouling, rotting, decay, disfigurement, toxification or weakening.
Catalyst, accelerator	Reagent added in very small amount to increase the rate of curing.
Cohesive failure	Failure of a joint within the bulk of the sealan.
Compatibility	The property of a sealant to remain in contact with another material without unfavourable physical or chemical interactions. Compatibility does not imply adhesion.
Consortium (consortia)	Blend/s of different micro-organisms in aqueous media to which sealants and joints are exposed.
Curative, curing agent	Chemical agent responsible for effecting the cure of a sealant.
Cure	Setting of the sealant, from the liquid to a solid state, by a chemical process such as polymerisation, chain extension or crosslinking.
Failure	The point at which the seal no longer performs the function for which it was installed.
Filler	Inert insoluble material dispersed in a polymer in the form of particles to cheapen, reinforce, colour or otherwise modify it.
Glass transition temperature	Middle of the temperature range at which a polymer changes from a glass to a rubber (Tg).
Hydrolysis	Breakdown of chemical groups in a substance by reaction with water.
Joint	Bonded assembly of two primed substrates with sealant between.
Joint filler	Compressible non-adhesive material used to fill movement joints during their construction.
Microbiological deterioration	Deterioration or degradation of materials as a consequence of microbiological activity.
Modulus	Intrinsic stiffness of a material.
One-part sealant	Sealant supplied ready for use.
Phase 1 absorption	Rapid absorption of water by a sealant, via a process of diffusion, until a saturated solution of water in sealant is achieved.

Phase 2 absorption	Absorption of water by diffusion into a polymer that accumulates by osmosis in little pockets or "clusters" around impurities in the sealant.
Plastic sealant	Sealant in which the stresses induced as a result of movement at a joint are relieved.
Plasto-elastic sealant	Sealant that has predominantly plastic properties with some elastic recovery when stressed for short periods.
Plasticiser	Material (usually an oil) added to reduce the glass transition temperature and/or the modulus of a polymer.
Polymer	Substance with molecules consisting of one or more structural units repeated any number of times.
Primer	Composition applied as a thin pre-treatment to the substrate to improve the adhesion of the sealant.
Retarder	Reagent added to reduce the rate of curing.
Sealant	Elastic or visco-elastic composition used to seal gaps between substrates subject to movements in service. One-part sealants cure by reaction with atmospheric moisture. Two-part sealants cure by the interaction of uncured sealant with a curative, the latter being added immediately before application.
Sorption	The uptake (*absorption*) or release (*desorption*) of a liquid by a polymer.
Substrate	Material of construction (in this case concrete) to which primer and then sealant are applied.
Two-part sealant	Sealant supplied as two separate components that have to be mixed before application.

Abbreviations

C1–C11	Microbial consortia (defined fully in A3.1)
PU1–PU9	Polyurethane sealants (explained and defined in A2.2.2 and A2.1.1)
PS1–PS5	Polysulfide sealants (explained and defined in A2.3.2 and A2.1.3)
Si1, Si2	Silicone sealants (explained and defined in A2.4.2 and A2.1.2)
Sew out	Sewage outfall (aerobic, submerged) at Severn-Trent Water, Finham
Sew 99%	99 per cent humidity, aerobic conditions over sewage at Severn-Trent Water, Finham
Sew anaer	Anaerobic sewage at Severn-Trent Water, Finham
Yorks pre	Raw, upland water before treatment, Yorkshire Water
Yorks post	Raw, upland water after treatment, Yorkshire Water
Tewkes pre	Pre-treated lowland River Severn water (submerged), near Tewkesbury
Tewkes post	99 per cent RH, post-treated lowland water, River Severn, near Tewkesbury
Soil	Soil (University of Warwick)

Notation

E	Extension of sealant expressed as a percentage of the original length
E_{max}	Mean E value at maximum force readings in tensile extension tests
ΔE_{max}	Percentage change in E_{max} relative to a control value
F	Force in newtons
F_{max}	Mean maximum value of F observed during replicate joint tensile extension tests
ΔF_{max}	Percentage change in F_{max} relative to a control value.
F_{25}	Mean value of F at 25 per cent extension during replicate joint tensile extension tests.
ΔF_{25}	Percentage change in F_{25} relative to a control value
σ	Tensile stress as force ÷ original cross-sectional area (MPa)
σ_{25}	Mean value of σ at 25 per cent extension in replicate strip tensile tests
$\Delta\sigma_{25}$	Percentage change in σ_{25} relative to a control value
$\sigma_{100}(max)$	Maximum value of stress at 100 per cent extension in strip tensile tests (MPa)
$\sigma_{100}(min)$	Minimum value of stress at 100 per cent extension in strip tensile tests (MPa)
l	Length of strip test specimen (mm)
Δl	Mean change in length of replicate strip test specimens after testing (mm)
Tg	The temperature at which a polymer changes from a rubber to a glass
S_{25}	Mean value of stress (MPa) at 25 per cent extension

Bibliography

BS 6093: 1993 *Code of practice for design of joints and jointing in building construction.*

BS 6213: 1982 *Guide to selection of constructional sealants.*

CIRIA SP80: 1991 *Manual of good practice in sealant applications.*

ISO 6927: 1991 *Sealants vocabulary.*

Guide to joint sealants for concrete structures; ACI 504R – 77, *ACI Manual of Concrete Practice.*

AUBREY, D W and BEECH, J C (1989)
"The influence of moisture on building joint sealants", *Building and Environment*, Vol 24, No 2, pp 179–190.

BEECH, J C (1985)
"Test methods for the movement capability of building sealants: the state of the art", *Rilem materials and Structures*, Vol 18, No 108, pp 473–482.

BEECH, J C and AUBREY, D W (1987)
Joint primers and sealants: performance between porous cladding, BRE Information Paper 9/87.

DAMUSIS, A (Ed) (1967)
Sealants. Reinhold Publishing Corporation.

FELDMAN, D (1989)
Polymeric building materials. Elsevier Applied Science.

KLOSOWSKI, J M (1989)
Sealants in construction. Marcel Dekker Inc.

MANSFIELD, C (1990)
"Tests for the water resistance of construction sealants", *Construction and building materials*, Vol 4, No 1, March, pp 37–42.

PANEK, J R and COOK, J P (1984)
Construction sealants and adhesives. John Wiley & Sons, 2nd edition.

SEAL, K J (1988)
"The biodeterioration and biodegradation of naturally occurring and synthetic plastic polymers", *Biodeterioration Abstracts*, Vol 2, No 4, pp 296–317.

WOLF, A T (1991)
"Movement capability of sealants", *Construction and Building Materials*, Vol 5, No 3, September, pp 127–134.

WOLF, A T (1994)
"Ageing resistance of building and construction sealants, Part 1", in *Durability of Building Sealants*, ed J C Beech and A T Wolf. E & FN Spon.

WOOLMAN, R and HUTCHISON, A (1994)
Resealing of buildings: a guide to good practice. Butterworth Heinemann, ISBN 0 7506 1859 0

1 Introduction

1.1 AIMS AND OBJECTIVES

The primary aims of this project were to:

- formulate a sealant test procedure to assess susceptibility to microbial degradation
- evaluate the impact of microbial activity on sealant performance, ie assess the susceptibility of specified sealant formulations to microbial degradation
- identify the active microbial species
- assess microbial degradation in selected environmental conditions and correlate these with defined microbial consortia
- from the data produce "Guidance Notes" as to the avoidance/minimisation of microbial degradation.

1.2 MICROBIAL BIOFILMS

All surfaces in an aqueous environment provide a suitable interface for microbial attachment and subsequent biofilm formation. Biofilms consist of an accumulation of cells, not necessarily uniform in time or space, immobilised at a substratum and frequently embedded in an organic polymer matrix of microbial origin. They are distinguished from suspended microbial growth systems primarily by the critical role of transport and transfer processes, which are generally rate-controlling within the biofilm.

The physical, chemical and biological properties of the biofilm are dependent on the particular environment in which the biofilm has accumulated. It is the physical and chemical components of the aquatic and substratum environment that influence the species composition of the biofilm. The predominant microbes, in turn, modify the micro-environment of the substratum in a manner specific to their metabolic activity and in general, biofilms are spatially heterogeneous. Interactions between the microbial cells in the biofilm with each other and the microenvironment, including the substratum, are complex and varied. The substratum properties will influence the rate and extent of adsorption of microbes and in many instances may influence their metabolic activity.

Biofilms are therefore complex, dynamic, multi-component accumulations of cells that have far reaching implications, both advantageous and disadvantageous. The recognised phases in biofilm development are shown diagrammatically in Figure 1.1. This report focuses on the degradative potential of biofilms with respect to polymeric sealants used in the construction industry.

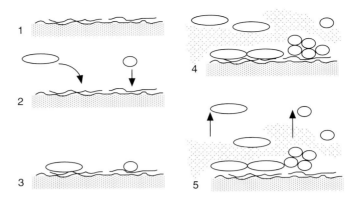

Figure 1.1 *Biofilm development phases. Key: 1. Preconditioning. Adsorption of macromolecules to the surface of the substratum. These macromolecules may serve as a nutrient source for the subsequent establishment of a biofilm. 2. Transport of microbial cells to the surface by diffusion, convection or active microbial movement. 3. Physical adsorption of cells to the surface which is reversible. 4. Growth, reproduction and formation of metabolic products including extra-cellular polymers leading to irreversible attachment. 5. Detachment due to erosion, sloughing and abrasion*

1.3 MICROBIAL DEGRADATION OF SEALANTS

Microbial degradation of sealants will by one of two methods, or a combination of both. The first is direct (active), degradation and subsequent utilisation of sealant constituents for growth. The second is by indirect physical/chemical effects upon the sealant as a consequence of microbial biofilm formation at the sealant/water interface. That is, the biofilm population derives no direct physiological advantage from the sealant *per se*.

Intensive studies on the biodegradation of synthetic polymers have clearly shown a close correlation between biodegradability and chemical structure. Synthetic polymers that biodegrade tend to have structures similar to those found in naturally occurring polymers. Data from such studies suggests that hydrolysis and oxidation are the primary processes responsible for polymer degradation. Among the important factors that contribute to biodegradability of polymers are the presence of hydrolyzable and oxidizable groups and the balance of hydrophobicity and hydrophilicity. In many instances a specified substratum may be inert to degradation by monocultures of microbes but readily susceptible to attack by symbiotic mixed populations. Such symbiotic associations are very important and ubiquitous in natural environments and are often cited as the cause of biodegradation of recalcitrant materials. It is probable that the polymeric sealants under study in this programme are more susceptible to degradation by such mixed populations that will predominate in the environments in question.

The pertinent interactions to be considered in the degradation of sealants are shown diagrammatically in Figure 1.2.

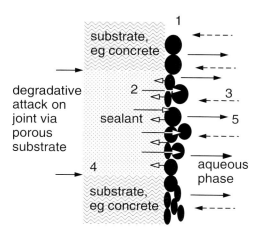

Figure 1.2 *Potential microbial interactions on a sealant/concrete joint in an aqueous environment. Key: 1. Attachment of microbes to the sealant/aqueous interface followed by degradative attack on the sealant by microbial enzymes/metabolites. 2. Utilisation of breakdown or "leached" materials from the sealant for microbial growth and/or release into the aqueous phase. 3. Supply of nutrients from the aqueous phase with secondary microbial activity leading to degradation of the sealant. 4. Degradation of the substrate/sealant joint by microbial intrusion. 5. Release of microbial cells or metabolites into the aqueous phase*

With respect to assessing the susceptibility for microbial degradation of a sealant formulation *per se* it is necessary to evaluate the metabolic potential offered to microbes by the individual components of the sealants as well as of the polymerised material. Identification of metabolisable sealant components will indicate the most likely biochemical degradative pathway(s) to be implicated. However, there is a considerable range of sealant formulations and a wide range of environmental conditions to which these may be exposed, thus presenting a broad spectrum for potential microbial activity and subsequent degradation. The situation is further complicated by the now widely recognised importance of mixed, ie multi-species, populations in achieving a degradative potential significantly greater than that which might be predicted from the metabolic activity of the component species. It must be recognised that many, if not the majority, of "real-life" degradative processes may be a consequence of mixed populations, the component species of which cannot be cultured in the laboratory as monocultures, nor effect the degradation alone.

A further complicating factor is the observation that, for example with polyethylene, the step that determines the rate of degradation is the rate at which molecular oxidation reduces the molecular weight of the polymer to a value suitable for microbial attack. The point of importance is that in many instances attack by micro-organisms may be a secondary process (Scott, 1975; Albertsson, 1993).

Published data on the microbial degradation of polymeric sealants is very limited; however, the following observations are pertinent to this programme.

Synthesized aliphatic polyesters are known to be susceptible to microbial degradation, particularly by fungi (Darby and Kaplan, 1968; Fields et al, 1974). Such isolates have been shown to assimilate a variety of aliphatic, but not alicyclic or aromatic, polyesters. It has also been reported that lipases from various microbes will hydrolyze several polyesters (Tokiwa and Suzuki, 1977; 1988).

Bacterial species capable of degrading polyethers have been isolated from soil and activated sludge (Kawai, 1987). Although originally thought to be via an oxidative process, anaerobic degradation has also been reported (Shink and Stieb, 1983; Dwyer and Tiedje, 1983; Grant and Payne, 1983).

With polyurethanes it has been shown that polyester polyurethanes are more prone to microbial attack than polyether polyurethane (Darby and Kaplan, 1968), the polyester segment of the structure being the initial site of attack (Seal and Pathirana, 1982).

From the published literature it is possible to draw the following conclusions:

- synthetic addition polymers with carbon chain backbones are more resistant to microbial degradation at molecular weights > 1000. Polyvinyl alcohol is an exception due to pendant hydroxyl groups, which are readily converted to hydrolyzable carbonyl groups
- synthetic addition polymers, such as polyacetals and polyesters, with hetero-atoms in their backbones, may biodegrade
- synthetic step-growth or condensation polymers are generally biodegradable to a greater or lesser extent depending on:
    - chain coupling (ester > ether > amide > urethane)
    - molecular weight (lower faster than higher)
    - morphology (amorphous faster than crystalline)
    - hardness (softer faster than harder)
    - hydrophilic (faster than hydrophobic).

The outline strategy for the selection of degradative microbes as used in this programme is presented in Figure 1.3.

The microbiological techniques that have been applied to elucidate the physiological activities and capabilities of biofilms are many and varied. However, there is no single procedure that is broadly applicable and none that lends itself to being an easy-to-use, reproducible assay. However, one of the primary, and most reliable, means of assessing the impact of a biofilm on a substratum is microscopy – light and scanning electron microscopy. Examination of the surface helps to establish the nature of the interaction between the biofilm and the substratum and yields data pertinent to the temporal build-up of the biofilm. In addition, it is of value to visualise the nature of the sealant surface as this may well play an important role in determining the extent of, or potential for, degradation.

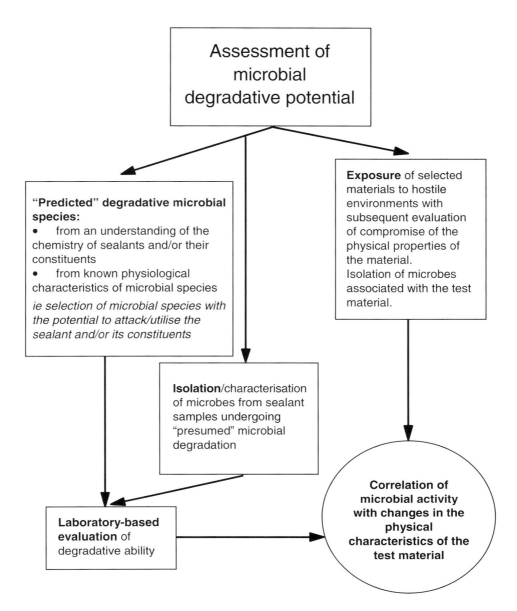

Figure 1.3 *Outline strategy for the selection of degradative microbes. There are three distinct routes to obtaining microbes with the capability of compromising the physical characteristics of a polymer. Two are laboratory-based – the isolation of micro-organisms from materials presumed to have undergone microbial degradation and secondly the selection of microbial species with the metabolic potential to utilise or compromise the sealant or its constituents. The third involves challenge by selected hostile environments. Subsequent correlation between the laboratory approach and the assessment of the environmental degradation will permit predictive laboratory based challenges to be undertaken in a more controlled manner than is possible by environmental challenge*

Distinction must also be drawn between the capacity to attach to sealants and the capacity to sustain growth via sealant breakdown. That is, consideration must be given to the possibility that the sealant may simply provide a suitable inert surface for attachment, microbial growth being at the expense of materials in the overlying fluid. In such circumstances, however, microbial products and the indirect effects of microbial growth may well have implications for sealant stability.

The major variables to be considered when addressing microbial degradation of sealants are therefore:

- the diversity and versatility of microbial metabolic activity, in particular the complexities of multi-species interactions
- environmental variability – the challenge environments range from nutrient poor (oligotrophic) treated potable water to raw sewage
- sealant composition (and potentially the conditions of cure).

In addition to the above, any effects by micro-organisms in aqueous media must be superimposed on those effects produced by water only. Thus, in looking for the effects of microbial attack, samples subjected to microbiological environments should ideally be compared against samples subjected to the same treatment in a sterile version of an otherwise identical environment.

Since micro-organisms cannot penetrate into the sealant bulk, it follows that any effects produced by microbial attack must, at least initially, be confined to the surface. It is possible that, at a later stage, some secretions or metabolic by-products may migrate into the bulk and produce noticeable changes. Within the time scale of the present study, microbial attack has been regarded as slowly progressing inwards from the surface, leaving the bulk unaffected. Clearly, under such conditions, any effects on an overall property such as modulus would become detectable earlier in thin sections, such as strips, than in thick sections, such as joints. Where a property depends strongly on surface conditions, however, such as breaking extension that involves crack propagation from surface flaws, the effects of microbial attack might become obvious at an early stage in both types of test piece.

As highlighted above, surface attack by micro-organisms might be expected to take one or more of several forms.

1. Physical modification of the surface by the attachment and build-up of a biofilm facilitating an accelerated rate of attack. It is possible that the biofilm would impart increased rigidity to the sealant that might produce noticeable modulus changes. It is also possible that this kind of effect is substantially independent of the generic type of sealant used.

2. Potential energy/sub-units for growth are either polymer or plasticiser. This would involve breakdown of these to smaller molecules, the most likely sites of attack being functional groups like ester, urethane, urea or disulfide. More unlikely would be the carbon-chlorine sites such as those in chloroparaffins. Breakdown of polymer in this way would produce an overall softened surface, perhaps with liberation of filler. Breakdown of the plasticiser would increase rigidity of the surface, an effect that might then be alleviated by migration to the surface of more plasticiser from the bulk.

3. Ancillary effects produced from microbial secretions or by-products of microbial attack, eg intermediate products of metabolism such as low-molecular-weight fatty acids could cause premature degradation of the polymer through the purely chemical action of hydrolysis. Concentration or pooling of such products near the surface would be assisted by entrapment in the biofilm and by a pitted or porous surface texture in the sealant.

To address the aims and objectives of this programme it was necessary to review the methods available to assess the mechanical performance of various sealant formulations and to establish microbiological and environmental challenge protocols. The selection of, and rationale behind, the test procedures adopted and evaluated are discussed in Section 3.

2 Sealant composition and microbial degradation

This section briefly reviews the chemical composition of the different sealant types and their expected performance when exposed to water and microbial challenge.

Some sealant formulations are commercially available, but others were simple generic formulations unlikely to be available, or suitable, for practical applications. These were included in the test programme to ascertain the susceptibility of individual sealant components to microbiological attack.

2.1 SEALANTS IN GENERAL

2.1.1 Types of materials present in sealants

Sealants are composed of several types of materials, as below.

Polymer   Although this is the component that receives most attention and is used to characterise the sealant type, it is not necessarily the material present in the greatest amount. The polymers of interest here are polyurethanes, polysulfides and silicones, although several others might have been used. All these polymers are initially viscous liquids that are converted into soft elastomers through a chemical curing reaction. The curing agent is either added as a second component (two-part sealants) or is moisture absorbed from the atmosphere (one-part sealants). Some cured sealants exist as a single phase of homogeneous material, whereas others comprise two or more phases of immiscible materials, although they are bound together chemically.

Filler   An inorganic material in the form of fine particles. It is added chiefly to modify rheological properties of the liquid sealant (eg to make it thixotropic), to reinforce the cured elastomer, and to extend – and therefore reduce the cost of – the whole sealant composition. It may also contribute important properties to the final cured sealant. It will modify such properties as water absorption and modulus. Fine fillers such as clays stiffen the sealant more than coarse fillers like ground chalk. The filler never dissolves in the sealant polymer and causes opacity of the sealant. Fillers are generally considered to be inert to oxidative ageing or heat degradation, but may be attacked by adverse corrosive environments such as acids.

Plasticiser   An organic oil that mixes with (ie dissolves in) the polymer and softens it. It may also reduce the Tg (the temperature at which the polymer changes from a rubber to a glass) and hence it improves low-temperature performance. Typical plasticiser oils for sealants are phthalate esters, phosphate esters and chlorinated paraffins. Although plasticisers are miscible in all proportions with uncured sealant polymers, they may be partially "squeezed out" (by the process of synersis) as the sealant cures. This limits the amount of plasticiser that may be used. The combination of softening by the addition of plasticiser and stiffening by the addition of filler may leave the sealant with a virtually unchanged modulus. However, these additions reduce its mechanical strength.

Other ingredients   Catalysts, pigments, coupling agents, antioxidants, rheology modifiers, surfactants etc may also be present, as required, but always in small proportions. Nevertheless, they may have disproportionately high microbiological activity, especially if heavy metals are present, and should therefore be given appropriate consideration.

2.1.2   Water sorption by sealants

A pure sealant polymer, when placed in water, will rapidly absorb water via a process of diffusion, until a saturated solution of water in sealant is achieved. This is usually at quite a low concentration, typically 1 per cent or less. The main effect of this absorption is plasticisation of the polymer (reduction of its Tg), reducing the modulus at a given temperature. The process is reversible, water being lost in dry conditions at a similar rate to restore the original properties of the polymer. A typical time of one day might be required for absorption or desorption of water by a pure sealant polymer bead to be regarded as essentially complete. In the initial stages the rate of absorption is proportional to the square of the time of immersion. This type of absorption is referred to later as *Phase 1* absorption.

Polymer-based products such as sealants are seldom pure enough for the above effect to be observed in an uncomplicated form. This is because there are inevitably hydrophilic impurities present (particles of adventitious inorganic matter, fillers, emulsifier residues, etc). The water diffusing into the polymer then accumulates by osmosis in little pockets or "clusters" at these impurities. These clusters enlarge slowly until the osmotic pressure is just counterbalanced by the elastic recovery forces of the polymer. If the polymer shows strong stress relaxation, the rate of absorption by this process will therefore be higher. The process results in a progressive change in optical refraction effects, leading to translucency and severe weakening of the polymer. The process is slow, the outer layers rapidly becoming saturated while the interior may remain much drier. It is referred to later as *Phase 2* absorption. Although the process of absorption may take many months to reach a certain overall level (say 10 per cent water), the drying-out process is much quicker, perhaps taking only one day. It is not easy to predict the extent to which the polymer will be left unaffected by this absorption/desorption cycle; certainly in the case of many joints the sealant is permanently weakened by the cycle.

The process of water absorption in elastomers generally has been studied by Briggs *et al* (1963) and concisely reviewed by Edwards (1985). The process of water absorption by polysulfide sealants has been studied by Hanhela *et al* (1986, 1986a). A study of the water absorption and performance of a range of sealant joints was described by Aubrey (1992).

The most accurate way of following the process of water sorption is by weighing the sealant at intervals while it is immersed. The results should be corrected for any materials lost from the sealant by leaching (next section). Changes in modulus that occur are large in the first phase of water sorption and less in the second stage.

2.1.3   Leaching of sealant components by water

Continued immersion in water will extract from the sealant any water-soluble components that can diffuse through the water-saturated sealant bulk. The effect will normally be restricted to removal of low-molecular-weight materials such as catalyst residues, dispersing aids like stearic acid etc. The process is very slow at ambient temperatures. Over a long period of immersion, only a small proportion of these will be lost, the amount depending on the surface area/bulk ratio of the sealant sample. In theory, however, even materials like ester plasticisers have a small solubility in water and will leach slowly, but the rate is usually negligible under test conditions.

The amount leached is readily obtained by the weight difference between the original dry sealant and the water-immersed, then dried, sealant. Samples should be given an identical conditioning treatment before and after leaching.

2.1.4 Hydrolysis processes in sealants

Polymers may also degrade by chemical processes of hydrolysis of groups such as ester, amide, urethane or urea groups. If these groups are part of the main chain the effect can be very serious, because the breaking of one group per chain immediately halves the average molecular weight of the polymer. Hydrolysis processes are slow at neutral pH but strongly catalysed by acid conditions and less strongly catalysed by alkaline conditions. Ether linkages are inert to hydrolysis, so the polyether-urethanes are more stable than polyester-urethanes to this kind of degradation. Polysulfides and silicones are generally less susceptible to hydrolysis than polyurethanes.

In the case of pure sealant polymers in pure water, the polymer will normally be water-saturated throughout its bulk so that degradation will be uniform through the volume of sealant. Commonly, however, the water will contain catalytic impurities that cannot penetrate into the polymer bulk, resulting in more degradation at the surface. The metabolism of micro-organisms might also produce pH changes highly concentrated at the surface. Hydrolysis of groups results in chain scission, causing the polymer to soften and eventually to become sticky and semi-liquid. Because of its strong acceleration by acids, care has to be taken when making sealant components (eg polyester polyols) to leave a minimum of free acid groups. The metal-containing catalysts, such as those used in various sealants in the present work, are also potential catalysts, and amine catalysts would be preferred from this point of view.

Little is known about the ability of micro-organisms to hydrolyse groups as part of their digestion processes, or to catalyse hydrolysis *via* waste products or pH changes. Nevertheless, it is often assumed that microbiological attack may well occur at hydrolysis-prone groups.

2.1.5 Modulus changes in sealants

It is common practice to measure a "secant modulus" (nominal stress for a particular extension) of sealants and sealant joints as an indication of their quality. This was done both on strips and on joints. It will be appreciated that all of the effects discussed above (especially curing, water sorption and microbial attack) will have their individual effects on modulus, and it may be very difficult to distinguish between them due to overlapping on a time-scale.

At this stage it is best to imagine the various changes in the form of diagrams for (a) a polymer that is fully cured before immersion (Figure 2.1), and (b) a moisture-curing polymer that is not fully cured before immersion (Figure 2.2).

The shapes and time-scales of the curves may vary enormously and will depend not only on the immersion conditions, cure rate etc but also on the generic type of sealant used. For the present, the curves are meant only to illustrate that different processes may be occurring, and the difference between the wet or dry moduli, with or without microbial attack, may vary widely depending on the time-point chosen.

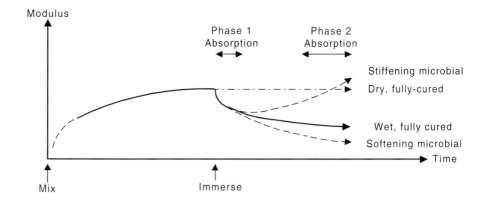

Figure 2.1 *Possible modulus change with curing, immersion and microbial attack, for a sealant fully cured before immersion*

The situation becomes more complicated if a sealant is immersed before curing is complete, particularly if it has a moisture-curing capability (Figure 2.2). In this case, the sealant will cure much more quickly when immersed. Consequently, the modulus will increase much faster and will be likely to exceed that of the dry-cured sealant over a range of times. The final modulus for the immersed sealant will be lower than that for the dry-cured sealant because of long-term (Phase 2) water absorption.

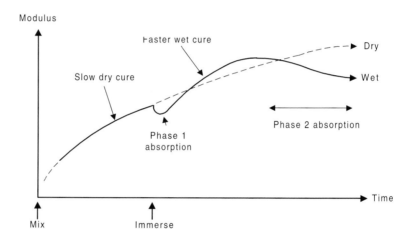

Figure 2.2 *Possible modulus change of a moisture-curable sealant when it is immersed before completion of dry-curing*

2.1.6 Breaking extension of sealants

It is also common practice to measure the breaking extension (percentage extension at maximum stress) for sealant joints as a measure of their performance. For the present work the extensibility of sealant strips has also been measured. A measure of the strength of the sealant in the form of the work or energy to failure may also be obtained as the force × distance moved by the jaws of the testing machine at failure. This latter has not been done in the present work.

Breaking measurements are not as reproducible as modulus measurements, as they depend strongly on the presence of stress concentrations that can initiate crack growth. Stress concentrations arise due to modulus differences between a soft sealant and a rigid substrate; thus, it is common to find failure occurring in joints adjacent to the concrete and, in the present work, between the sealant strip and the jaw of the testing machine.

Further, the crack will initiate from a pre-existing microscopic flaw or defect at the surface, so that the sharpness of the cutting tool or minor surface abrasions can have a large effect on strength. To overcome these effects in polymers generally, dumbbell-shaped test specimens are used, where the polymer is either moulded in situ or is cut from a sheet with specially designed sharp tools. Such procedures were considered impracticable for our tests, so that additional variability must be expected.

If this variability is overcome, the strength of an elastomer is found to depend on the sharpness of the crack tip. If processes such as crystallisation, filler hysteresis or stress relaxation blunt the crack tip, the polymer will be stronger since less of the input energy will be available for generating new surfaces. Where such effects do not arise, the tear is "brittle" (fracture surfaces have no residual deformation and can be fitted together again). Where there is energy dissipation, the tear is described as "knotty" (fracture surfaces remain rough and the crack clearly had a tortuous path). The polyurethanes of the present work fail in a "brittle" manner, whereas the polysulfides show more knotty tearing due to their higher stress relaxation.

The swelling of a sealant by water will make crack propagation easier since the reduction in T_g will reduce the hysteresis loss at the crack tip (at the same rate and temperature of test). Thus, a water-swollen sealant will be weaker than an unswollen one. The inverse will apply when a water-swollen sealant is dried.

The effect of microbial attack on the surface can conceivably make crack initiation easier by the introduction of flaws, particularly if the attack results in cracking or crazing of the surface. Under these circumstances a reduced breaking extension would be expected.

2.1.7 Adhesion of sealants

Retention of good adhesion is a prime requirement for all applications of joint sealants. However, poor adhesive performance is frequently responsible for premature failure. Surveys in both the UK and Japan, for example, have highlighted the predominance of adhesive failures on building facades (Woolman and Hutchinson, 1994).

In some other applications, there is a greater tendency for degradation and cohesive failure of the sealant itself to occur. When joint surfaces are subjected to traffic or water flow, for example, mechanical breakdown due to abrasive/erosive action may become very significant. Similarly, certain properties of the bulk sealant become very important where there is exposure to aggressive chemicals (including water) or a microbiologically active environment. However, while the response of the sealant itself to severe service conditions demands considerable attention, a focus upon the adhesive requirements must be maintained, particularly as good initial adhesion does not guarantee acceptable long-term service performance.

An early report on the failure of joint sealants in sewage installations (Appleton, 1973) placed some emphasis on the lack of resistance of various sealants to biodegradation. However, this report also notes that, in general, the initial mode of failure was loss of adhesion, in some instances after only a few weeks exposure. In these and similar environments, increased resistance to microbial attack is unlikely to give improved long-term adhesion.

Factors affecting adhesion

Many factors contribute to adhesive performance and most are critical as each can cause premature failure. These factors may be grouped under three main headings:

- design
- material properties
- workmanship.

As the present research is most immediately concerned with the evaluation of material properties, the influence of design and workmanship on adhesive performance will not be discussed (but see, for example, BS6093, 1993; BS6213, 1982; CIRIA, 1991; Ledbetter, Hurley and Sheehan, 1998; the Bibliography).

The properties of a sealant system that control adhesion can be divided into two distinct (although interactive) areas.

1. The inherent ability of a given sealant/primer system to form and maintain an adequate bond at the substrate surface.

2. Those properties of the bulk sealant that determine the stress placed on the interface as a result of joint movement.

In the latter case, the stiffness of the sealant and its ability to undergo stress relaxation are of main concern.

In contrast to the more plastic sealants, highly elastic products generally exhibit relatively little relaxation and any stress imposed on the bondline tends to be maintained. Consequently, good adhesion is essential.

Even in the absence of relaxation effects, interfacial stresses cannot be calculated directly from the modulus of the sealant as, due to restraint by the substrate, stress concentrations arise. The sealant supplier, therefore, recommends an acceptable level of joint movement (implicitly the acceptable stress), which is based on tests carried out with bonded rather than isolated sealant specimens. For high-performance sealants, the recommended strain limit for a sealed joint is generally at least an order of magnitude lower than the strain that can be accommodated by the sealant itself.

Failure mechanisms

If excessive strain is applied to a bonded test specimen, failure may take place cohesively within the sealant (or substrate in some cases), adhesively (ie a clean separation of components at an interface) or by a combination of these mechanisms. The former mode of failure is generally to be preferred, as the full potential of the sealant is then realised. However, adhesive failure may be acceptable, particularly with sealants that undergo stress relaxation, provided that a satisfactory degree of extension is achieved beforehand.

Adhesion under wet conditions

When sealant systems are intended for use in wet service environments, bonded test specimens should ideally exhibit the same type of failure after water immersion as shown by dry controls. A similar, or greater, level of extension should also be found, although this should not result from excessive softening of the bulk sealant. The longer the immersion period for which the sealant system conforms to this behaviour, the greater the confidence that performance in service will be satisfactory.

Wet conditions can lead to a change from cohesive to adhesive failure for two main reasons.

1. The sealant initially becomes softer and then becomes stiffer (Figures 2.1 and 2.2) and a greater stress is then placed on the bondline at a given strain.

2. An interface is downgraded directly by the action of water that has diffused to the bondline. Such action may include displacement of a primer or sealant by a predominantly physical process (to achieve a more favourable thermodynamic state) and the disruption of chemical bonds (hydrolysis). Primers and surface conditioners can be used both to maximise the bond strength attained at the sealant/substrate interface and to resist the physical/chemical processes noted above.

As many sealant applications involve exposure to moisture, adhesion testing is commonly carried out on proprietary products. A recent review of the ageing resistance of joint sealants has included a general discussion of the effects of water (Wolf, 1996). More detailed assessments of the performance of civil engineering sealants under wet conditions have been carried out in earlier projects (CIRIA, 1987; CIRIA, 1992; Anon, 1988) and complementary research at the Building Research Establishment (Beech and Aubrey, 1987, Aubrey and Beech, 1989; Mansfield, 1990). This work included development of the water-resistance test, published as a Water Industry Specification (WIS 4-60-01) and described in Section 3.4.2 of this report, and a detailed examination of the factors that can affect joint performance at the sealant/substrate (concrete) interface, particularly the presence of suitable primers.

2.2 POLYURETHANE SEALANTS

2.2.1 Polyurethanes in general

Polyurethanes are the materials formed by the step-growth reaction of polyhydroxy compounds with polyisocyanates. The basic reaction involved in their formation is the addition:

$$\text{—OH} \quad + \quad \text{OCN—} \quad \rightarrow \quad \text{—O-CO-NH—}$$
$$\text{hydroxyl} \qquad \text{isocyanate} \qquad\qquad \text{urethane}$$

Also often important is the addition of amino groups:

$$\text{—NH}_2 \quad + \quad \text{OCN—} \quad \rightarrow \quad \text{—NH-CO-NH—}$$
$$\text{amino} \qquad\qquad\qquad\qquad\qquad \text{substituted urea}$$

and water:

$$\text{H}_2\text{O} \quad + \quad 2\,\text{OCN—} \quad \rightarrow \quad \text{—NH-CO-NH—} \quad + \quad \text{CO}_2$$
$$\qquad\qquad\qquad\qquad\qquad \text{substituted urea}$$

Note that the reaction with water joins together two isocyanate groups and forms a molecule of CO_2 gas. It can therefore be used in polymer-building and foam-formation simultaneously. All of these reactions are slow, but can be catalysed by either tertiary amines or organometallic compounds (eg dibutyl tin dilaurate).

Some of the polyhydroxy compounds ("polyols") are themselves polymers of low degree of polymerisation (molecular weight typically about 3000–4000), usually either polyethers (formed by chain addition reactions) or polyesters (formed by step-growth polycondensations). Because of this, although urethane formation is an important polymer-forming reaction, urethane groups are not necessarily present in high concentration in the final polymer. Usually the predominant group is either ether or ester, the final polymer often being referred to as polyether-urethane or polyesterurethane. Sometimes the number of urethane groups present is extremely small and is greatly exceeded by substituted urea groups (see above) as well as by ether (or ester) groups. In this case the product is often referred to as a polyether-urea or a polyesterurea, or sometimes simply as a polyurea.

Very small amounts of carboxylic acid groups may have a disproportionately large effect by catalysing the hydrolytic degradation of polyester-urethanes. The presence of small amounts of unsaturation may be a significant factor in determining the resistance of polyether-urethanes to microbial degradation.

An important feature of polyurethanes is their great versatility. By suitable adjustment of the formulation, products suitable for a wide range of applications may be made, including:

- vulcanizable rubbers (for conventional processing)
- thermoplastic rubbers (for injection moulding or solution processing)
- reaction injection mouldable (RIM and RRIM) rubbers or resins
- castable rubbers or resins
- varnishes and other surface coatings
- adhesives of both contact and reactive types
- flexible and rigid foams
- elastic fibres
- sealants.

To understand the rationale governing the formulation of polyurethane compositions for the various applications, it is important to recognise from the outset the following two features of polyurethanes.

(a) A polyurethane is normally made from the reaction together of three essential types of raw material: isocyanate, polyol and chain extender. (The latter is so-called because it is a small molecule that can link the ends of low-molecular-weight polymer molecules to form very high-molecular-weight polymer molecules.) A technologist formulating a polyurethane must be aware of the contribution of each to the polymer structure. The most common isocyanate used in most of the above products (including sealants) is 4,4'-diphenylmethane di-isocyanate (MDI) in various degrees of purity. The polyol is the major component and is therefore most important in determining physical properties of the polyurethane. The chain extender is a small molecule such as butane diol or water, but simple diamines are also used sometimes. The chain extender may be called something different, eg "curing agent", but it is the function rather than the name that is important and must be recognised. In moisture-curing products such as varnishes or sealants the water is not deliberately added but diffuses into the polyol/isocyanate blend from the atmosphere. If this process occurs slowly, the CO_2 liberated can diffuse away from the polyurethane without causing foaming. Such products, of course, cure from the outside inwards and the inside may remain undercured for a very long time. Note that catalysts, colours, flame retardants, blowing agents, fillers etc may also be present, but most will not normally react to form part of the polymer structure.

(b) In most cases the polyurethane polymer is not a simple, homogeneous (single-phase) material. It consists of minute regions, referred to as domains, of "hard" and "soft" material. These domains can only be seen under the high magnification of electron microscopy. Separation of the initially homogeneous mixture into these phases occurs at an early stage of reaction, and occurs because the polymer formed is a block copolymer (also called a segmented copolymer). The polymer chains consist of high-Tg, "hard" blocks or segments (made from isocyanate and chain extender) and low-Tg, "soft" blocks or segments (made from polyol), which segregate to form the "hard" and "soft" domains. This resulting two-phase morphology is very important in determining the toughness and strength of the product. Although it is possible to make polyurethanes without this two-phase morphology, they are seldom encountered in practice because of their poor mechanical properties. The model polyurethane sealants used in the CIRIA study are one of these exceptions, showing only one phase but poor mechanical strength (see below).

The proportions of hard and soft blocks can readily be varied by adjustment of the proportions of isocyanate, polyol and chain extender. When the hard block content is in excess, it will form the continuous phase and the polymer will take the form of a rigid plastic toughened by the presence of the soft rubbery inclusions. When the soft block content is in excess, the polymer will be a rubber, reinforced by the hard domains.

2.2.2 Polyurethane sealants used in the present work

The polyurethane formulations used have been designated as PU1 to PU9 (Appendix A2). Perhaps the formulation with the most "standard" ingredients is PU3, and this will be discussed first. It consists of:

	Pts wt
Polyether polyol	22.2
Isocyanate	1.6
Chlorinated plasticiser	26.0
China clay filler	13.0
Dibutyl tin dilaurate catalyst	0.05

The above are mixed and allowed to react to form the final polyurethane. Since the polyether polyol requires 1.5 parts of the isocyanate for full reaction, this leaves an excess of 0.1 parts, which will form rather unstable allophanate groups and eventually react with atmospheric moisture. A simple calculation shows that the amount of water required to neutralise this remainder is negligible (0.007 parts). Thus, if allowed to cure properly, the hard segment content of the polyurethane formed is negligible, and the polymer must consist essentially of only a single phase. This will give the maximum softness (low modulus) but low strength, as has been observed.

Furthermore, it can be calculated that the polymer itself will contain one mole of ether groups per 54 g, and one mole of urethane groups per 2136 g. Thus, the ether groups outnumber urethane groups by about 40 to 1.

The components of the final PU3 sealant will therefore be:

Polymer (single-phase polyether-urethane)	100 parts
Plasticiser (presumably dissolved in the polymer)	109 parts
Clay filler (as dispersed particles)	55 parts
Tin catalyst residue	0.2 parts

The remaining PU sealants differ from this composition as follows:

PU1   A polyester polyol is used instead of the polyether one, so that ester groups will replace the ether ones in numbers that cannot be calculated from the manufacturer's information. Generally, a polyester-urethane shows less water absorption than a polyether-urethane, but it also is more likely to hydrolyse.

PU2   As PU3 except that an aliphatic isocyanate has been used in place of Desmodur VLR20. This should give better colour and hydrolysis resistance.

PU4   As PU3 but an amine catalyst replaces the tin one. Perhaps poorer resistance to microbial attack might be expected.

PU5   As PU3 but a mercury catalyst replaces the tin one. Perhaps better resistance to attack should be expected.

PU6   The chlorinated hydrocarbon plasticiser oil is replaced by a phthalate ester oil (2-ethyl hexyl benzyl phthalate). The general view was that this should result in inferior resistance to attack.

PU7   The clay filler is omitted, otherwise it is as PU3.

PU8   The clay filler, plasticiser and catalyst are all omitted, leaving a very simple formula consisting only of polyether-urethane.

PU9   Clay filler, plasticiser and catalyst are omitted, and the polyester polyol replaces the polyether one of PU3. The product consists only of polyester-urethane.

It should be pointed out that there is no positive or well-documented information to support the views expressed under PU2–PU6, above.

Even shorter abbreviations will be used in further discussion, for example they will simply be referred to as **PU1** (polyester), **PU2** (aliphatic), **PU3** (standard), **PU4** (amine catalyst), **PU5** (mercury catalyst), **PU6** (phthalate plasticiser), **PU7** (no clay), **PU8** (polymer only – polyether) and **PU9** (polymer only – polyester). The words in brackets merely refer to the major change in composition from the standard grade, **PU3**.

2.3   POLYSULFIDE SEALANTS

2.3.1   Polysulfides in general

Polysulfides differ from polyurethanes in that they show a much more limited versatility, being mainly confined to the field of sealants.

The type of polysulfide of interest here is based on the condensation products of bis (2-chloroethyl) formal with sodium polysulfide. The polymer obtained by this condensation has disulfide linkages that may be reductively cleaved by sodium hydrosulfide and sodium sulfite in different ratios to form a range of liquid, low-molecular-weight polysulfides (MW range about 1000 to 8000). The general formula for such a polymer would be:

$$HS-[-CH_2-CH_2-O-CH_2-O-CH_2-CH_2-S-S-]_n-CH_2-CH_2-O-CH_2-O-CH_2-CH_2-SH$$

A typical average value of n for a well-known liquid polymer is 23. The curing of these liquids is effected by oxidation of the terminal -SH groups to form disulfides, and hence a high molecular weight polymer of the simple structure:

$$-[-CH_2-CH_2-O-CH_2-O-CH_2-CH_2-S-S-]_n-$$

where n is now a very large number.

It can be seen that this polymer has two ether groups for every disulfide group and might therefore be better referred to as a polyether-sulfide (as was the case with the polyurethanes). In this case, the ether groups have different reactivity as they are in the acetal (-O-CH$_2$-O-) form, being particularly susceptible to oxidation.

The polymer as shown above is linear (uncross-linked), but in practice a small amount of cross-linking would be present, provided by the introduction of a little trifunctional halide (eg trichloropropane) in the original reaction mix. Alternatively, a branched liquid polymer, prepared in this way, may be blended with a linear liquid polymer in order to obtain a cross-linked product. Thus, gel formation occurs when the liquid polymer blend is cured, at a stage of reaction depending on how much trifunctional component is present. The product used in the present work is probably a blend of this type.

The mechanism of curing of polysulfides is not at all straightforward, and the properties of the final polymer can depend strongly on the type of oxidant used. Four types of oxidant in common use as curing agents are manganese dioxide (MnO$_2$), lead dioxide (PbO$_2$), calcium peroxide (CaO$_2$) and sodium dichromate (Na$_2$Cr$_2$O$_7$) The reactivity of these curing agents is generally regarded to be in the order:

PbO$_2$ > Na$_2$Cr$_2$O$_7$ > MnO$_2$ > CaO$_2$

However, each of these curing agents is a complex system in practice. For example, MnO$_2$ powder alone is a very poor curative, but when "activated" by heating with alkali (to introduce manganate or permanganate ions) and when used in the presence of moisture or plasticisers is much more active.

Cured polysulfides made using these oxidants show different amounts of various forms of cross-link bonds. Some bonds are covalent and very durable, others are hydrogen bonds or adsorptive bonds with filler, both of which can be destroyed by moisture, especially at higher temperatures, or in polar media. A comprehensive comparison of the durability of manganese- and dichromate-cured sealants has been carried out by Hanhela *et al* (1986).

In the present study we have a comparison of otherwise identical manganese and lead-cured products, as well a very simple, unfilled, polymer cured with zinc peroxide.

2.3.2 Polysulfide sealants used in the present work

Five types of polysulfide sealant were used, denoted by PS1–PS5. The full formulation for each of these is shown in Appendix A2, but some explanation of the functions of the different components is given below.

The formulations PS1–PS4 are all based on a liquid polymer that is a polysulfide of molecular weight 4000, containing 0.5 per cent of a polyfunctional material for cross-linking. They also contain the inorganic filler titanium dioxide, stearic-acid-coated, precipitated calcium carbonate and stearic-acid-coated, ground calcium carbonate. The titanium dioxide is an opacifying pigment; the precipitated calcium carbonate is a fine grade producing some reinforcement (and hence stiffening) of the sealant; the ground calcium carbonate is a cheaper grade with coarser particles and showing little stiffening. The stearic acid coating helps to disperse these fillers in the polymer, and a little extra is present in PS2 and PS3, possibly to help control the cure rate (see below). All four formulations also contained a little sulphur (of uncertain function) and a little hydrocarbon tackifier resin as an adhesion promoter.

The major differences between the three formulations PS1–PS3 were in the type of plasticiser and the type of curing system used. PS1 and PS2 used a chlorinated paraffin plasticiser, whereas PS3 used octyl benzyl phthalate. PSI used manganese dioxide as the curative, whereas PS2 and PS3 used a lead dioxide curative. In each case the curative was used as a dispersion in the appropriate plasticiser. For the MnO_2 cure, a small amount of tetramethyl thiuram disulfide was added to the curative as an accelerator, but for the PbO_2 cure, a small addition of stearic acid was added as a retarder. Both the manganese dioxide and lead dioxide curatives were of fine particle size; the manganese dioxide had also been specially activated by alkali treatment.

Polysulfide PS4 was identical to PS1 except that it contained a small amount of a thixotrope, so as to make the formulation more closely resemble a common type of industrial sealant.

Polysulfide PS5 was intended to be as simple a formula as possible for a polysulfide, containing only polymer and a blend of zinc and calcium peroxide as curative. This was prepared by conventional rubber processing methods and was received as a cured sheet. It could not, therefore, be used for experiments in joints.

The main features of each of the polysulfide polymers may be abbreviated as follows:

- PS1 (manganese cure, chlorinated plasticiser)
- PS2 (lead cure, chlorinated plasticiser)
- PS3 (lead cure, phthalate plasticiser)
- PS4 (manganese cure, chlorinated plasticiser, thixotropic)
- PS5 (simple polysulfide).

From the point of view of microbial attack, it was considered from earlier observations (Morton International Polymer Systems) that resistance to attack should be in the order:

PS1, PS4   >   PS2   >   PS3.

The possible position of PS5 could not be suggested, due to lack of information.

2.4 SILICONE SEALANTS

2.4.1 Silicones in general

The description "silicone" is applied to those polymers that have a siloxane main chain of alternating Si and O atoms:

$$-\{-\underset{R}{\overset{R}{Si}} - O-\}_n-$$

In silicone elastomers the majority of the groups R are methyl ($-CH_3$), although other groups such as phenyl ($-C_6H_6$) or trifluoropropyl ($-C_3H_4F_3$) may be present for special physical properties. Smaller amounts of other groups may also be present for reactive functionality such as for cross-linking (see below). In addition to the silicone polymer, commercial silicone elastomers often also contain fillers for reinforcement, pigmentation, cheapness or other purposes. Smaller amounts of other additives, including catalysts and curing agents, may also be present.

The silicone elastomers are derived essentially from hydrolysis products of the dichloride:

$$Cl\text{-}Si(CH_3)_2\text{-}Cl \xrightarrow[\text{(fast)}]{H_2O} HO\text{-}Si(CH_3)_2\text{-}OH \xrightarrow{\text{(spontaneous)}} -(-Si(CH_3)_2\text{-}O-)_n-$$
$$\text{silicone}$$

The silicone product obtained by this scheme contains a high proportion of cyclic oligomers, which are converted into linear form by equilibration at high temperature. This involves heating with alkali at 150–200 °C. At the same time, monofunctional groups are added to control chain length and reactive groups are added for subsequent cross-linking. Two main types of room-temperature-vulcanizing (RTV) liquid polymers are made by this process.

RTV polymer for two-part systems:

- n = (typically) 500–1500
- functional groups such as alkoxy, acetoyl, hydride, vinyl are incorporated
- viscous liquid consistency
- used in encapsulation, moulding, insulation, coatings and flexible moulds
- (Note: different functional groups will be present in the two parts. These react with each other during the cure. The cure temperature may be from room temperature up to about 100 °C.)

RTV polymer for one-part, moisture-curing systems:

- n = (typically) 200–1500
- moisture-reactive functional groups such as alkoxy, acetoxy, benzamido, alkanoneoximino are incorporated. The two latter types are referred to as "benzamide" and "oxime" curing systems, respectively
- viscous liquid consistency
- used in adhesives, sealants, coatings.

In the case of silicone sealants, the one-part systems are of most interest in the present work due to their simplicity in use. Thus examples of both the oxime and benzamide curing systems were used. In both cases the curing reaction needs to be catalysed by the use of organotin compounds.

The properties of silicones generally are dominated by the unusual structural features of heat-resistant bonds, flexible Si-O main chain bonds, and close packing of methyl groups. These impart the distinctive properties of high heat resistance (eg a useful life of up to two years at 200 °C), high flexibility to low temperatures (Tg -123 ± 5 °C), and low surface energy (low intermolecular attractive forces). The polymers also show good resistance to water, UV and oxidation. The main disadvantages are the severe swelling in petrol, lubricating oils and diesel fuels, the severe corrosive effects of strong acids and bases, and the low strength of the rubbers. The strength can be increased by the use of reinforcing fillers such as fumed silica, but even then does not approach the strength of most industrial rubbers. The permeability of silicone rubbers to gases and compatible liquids is very high compared with other rubbers, usually about 10–20 times higher. Thus, oxygen, carbon dioxide and water diffuse at very high rates through silicone rubber sheets. This is clearly an attractive feature in the case of one-part silicone sealants, enabling them to cure quickly.

2.4.2 Silicone sealants used in the present work

The silicone sealants used in this work are of two basic types, being representative of industrial products (Appendix A2).

The silicones denoted by Si1 and Si2 are oxime-cured. The main difference between them is that Si2 contains calcium carbonate filler in addition to fumed silica reinforcing filler, which is present in both. Both contain three types of low-MW silicone polymer:

- a reactive OH-terminated polydimethyl siloxane (the main curing polymer)
- an unreactive polydimethyl siloxane oil used as a softener
- an aminofunctional siloxane used as an adhesion promoter.

In addition to these, a combination of tri- and tetra-functional butanoneoximinosilanes is present in each as the moisture-sensitive cross-linker. An organotin carboxylate is present as catalyst in both cases, and an ethylene/propylene oxide copolymer is used as stabiliser in the case of Si2.

The curing reaction involves displacement of the butanone oxime from the silane by water:

$$- Si - O - N = C (CH_3)(C_2H_5) + H_2O \rightarrow - Si - OH + HO- N = C (CH_3)(C_2H_5)$$

The silanol groups so formed then condense spontaneously with the silanol groups in the liquid polymer to form a cross-linked network. The butanone oxime liberated is a rather involatile material and takes some time to diffuse away from the sealant. It is a powerful biocide and is therefore expected to protect the sealant from attack for a short time after curing (eg a week or two in a typical joint).

The second type of silicone sealant (Si3) used in this work is a benzamide-cured product from a different manufacturer. This sealant is similar in that it contains the hydroxy siloxane, the dimethyl siloxane, the calcium carbonate and fumed silica fillers, and the organotin carboxylate catalyst. The exact characteristics of these materials and their amounts will not, however, be the same. There is also a small amount of titanium dioxide and carbon black for coloration purposes, which may be ignored. There is no stabilising copolymer and no adhesion promotor in Si3.

The main features that differ in Si3 are (i) the presence of a low proportion of a titanium ester compound, which is said to produce some stress relaxation in the final sealant, and (ii) an arylamido silane material as cross-linker. It is presumed that the arylamido material is hydrolysed by atmospheric water to a silanol in a manner analogous to that of the oximino compound, above, resulting again in silanol groups that spontaneously condense. The final cured sealant in this case is likely therefore to contain some non-volatile by-product such as benzamide.

The sealants will be referred to below as Si1 (oxime-cured), Si2 (oxime-cured, filled) and Si3 (benzamide-cured, filled).

3 Test procedures

This section describes the principles and rationale underlying the procedures, which have been examined, and eventually tested, to meet the objectives of the programme as laid out in Section 1.1. The detailed methods used and results obtained are presented in full in the Appendices.

Sealant strips and sealant joints were placed in selected environments for defined times and the extent of microbial degradation of the polymer ascertained via changes in its mechanical properties. In addition, the sealant strips were examined by microscopy and the extent of surface change categorised in comparison to control samples.

In an attempt to standardise the microbiological challenge imparted by a particular environment the sealant strips were exposed to defined microbial consortia. These were prepared, as far as possible, to mimic the biochemical/chemical capabilities of the microbes known to be indigenous to each of the selected environments.

3.1 TEST SPECIMEN GEOMETRY AND PREPARATION

3.1.1 Test specimens for evaluation of the performance of sealant materials in aqueous environments

Published work on the biodegradation of polymers generally uses test-pieces of no specified shape and size, since the microbial effects are evaluated by a visual or microscopic examination of the surface. Sometimes this involves examination of sealants in a practical environmental situation, such as the study described by Appleton (1973) on tests covering a range of sealant types. Sometimes the tests involve exposure of polymers to laboratory cultures of micro-organisms, such as the tests described by Kaplan *et al* (1968) in which they attempt to determine the ability of various polyurethanes and their individual components to sustain fungal growth.

It was decided that the test specimens to be used in this present study would need to be suitable for both laboratory and environmental tests, with evaluation of microbial attack by light microscopy and scanning electron microscopy (SEM). Furthermore, it was decided that the test specimens should be sturdy enough to conduct mechanical measurements of modulus and strength, tests that apparently have not previously been carried out in studies of the microbiological deterioration of sealants.

For these mechanical tests, the test specimens would need to be in the form of thin sheets to maximise the surface area and so maximise the potential for microbiological attack, producing detectable mechanical changes after a reasonably short exposure time. Consideration was given to three kinds of test specimen:

1. Cast sheets of sealant of sufficient dimensions to allow the determination of fracture energy by measuring crack propagation rate under a fixed load in pure shear.

2. Dumb-bell test specimens of standard dimensions cut from cast sheets. These would allow modulus and stress-strain behaviour to breaking point to be obtained by standard procedures. Some such specimens were made and tested.

3. Simple rectangular specimens cut from cast sheets. These would be quick and easy to prepare and would be suitable for approximate modulus measurements, but the breaking strength and extension were expected to be more variable.

After much consideration, it was decided to opt for test pieces of Type 3. The rationale behind this choice was that the test pieces could be made and tested with a minimum amount of expertise, and they could be produced rapidly and in very large numbers, thus allowing many variables and replicates to be tested. Furthermore, it would be easy to compute the amount of swelling or leaching from dimensional and weight changes in such specimens, which would provide useful ancillary information.

The sealant strips were prepared as described in Appendix A1. Sealant formulations are shown in Appendix A2. There were nine polyurethane, three silicone and five polysulphide formulations.

3.1.2 Tests to evaluate the performance of sealant joints in aqueous environments

CIRIA has previously addressed the performance of sealant joints in aqueous environments (RP355 Phases l and 2, 1987–1992) (see Section 2.1.7). A Water Industry Specification (1991) has been published and includes the test method used in that work. In these studies, attention was focused on the absorption of water, the diffusion of liquid water to the sealant/primer/substrate interfaces, and the requirement for the sealant to be fully cured. It was decided that it would be logical to use the same test joint and test procedure in the present work (see Appendix A1.2).

3.2 PHYSICAL TEST METHODS

3.2.1 Testing of strips

The stress-strain behaviour to breaking point of all sealant strips was tested wet at 50 mm/min^{-1} and room temperature using, by courtesy of Morton International, a Davenport-Nene testing machine. All sealant strips were tested wet except for the dry control. For the test, a standard 10 mm length at the top and bottom of the strip was secured in smooth calipers and positioned on the instrument. This left a standard 80 mm of specimen length to be strained during the test. The rate of 50 mm/min^{-1} was chosen so as to give approximately the same basic rate of straining of the sealant as in the joints (3.2.2 below). Force-extension data were obtained for each specimen and this was converted into stress-extension data after measuring the test specimen thickness. From these results the individual values of stress (MPa) at 25 per cent extension (s_{25}) and the percentage extension at maximum stress (E_{max}) were noted. It was observed that breakage of the strip invariably occurred proximal to the clamping point where the sealant had been compressed.

Triplicate tests were carried out for each nominally identical sealant and exposure condition, and the results averaged (tabulated in Appendix A4). Also recorded in these tables are the average change in length due to immersion and maximum and minimum individual values of stress at 100 per cent extension (σ_{100} (max) and σ_{100} (min)).

In addition to the above tests, the initial and final weights, thicknesses and lengths were measured and recorded.

Stereoscanning electron microscopy (SEM) and light microscopy were carried out on those samples showing significant change.

3.2.2 Testing of joints

The test method for joints followed that required by WIS 4-60-01 (March 1991: Issue 1) and is given in detail in Appendix Al.

3.3 MICROBIOLOGICAL TEST METHODS

While joints or strips can be exposed to specific environmental challenges this has severe limitations with respect to reproducibility and objectivity. To overcome these, microbial consortia, or more correctly combinations of species, that reflect the perceived primary microbial activities of the environments of interest were assembled for use in the controlled laboratory tests (Appendix A3) and used to challenge various sealant test strip formulations.

3.3.1 Microbial consortia

Microbial species were selected on the following basis:

- from culture collection isolates previously implicated in polymer degradation
- those microbes with metabolic capabilities compatible with potential utilisation of the polymers (and their individual constituents) as assessed from the chemical structure of the polymers
- those microbes isolates from suspected incidences of polymer degradation (Section 4.1).

As discussed in the Introduction, it is not possible, in general, to formulate consortia that adequately reflect the microbial activity of a specified environmental situation. Although the importance of mixed culture interactions in the environment is widely recognised, such consortia and their interactions are both poorly characterised and poorly understood. Each consortium used in this programme was an assemblage of microbes broadly having the same biochemical capabilities – and hence degradative potential – as those known to be indigenous to each of the selected environments. That is, each microbial species was predicted to be of primary importance in a specified environment.

Appendix A3 lists the consortia used in this work.

3.3.2 Microbial consortia: challenge test

The selected microbial species were grown as monocultures in their respective growth medium until the late exponential/early stationary phase of growth, ie population approaching nutrient limitation. A consortium was prepared by mixing 5 ml aliquots of each monoculture. Sealant strips, following weight and dimensional measurements, were immersed in 3 ppm hypochlorite for 10 minutes, subsequently washed with sterile distilled water and placed in sterile tissue culture bottles (Figure 3.1). To these were added 50 ml of sterile distilled water (pH 6.8), 0.5 ml of a minimal salts medium and 5 ml inoculum. Incubation was either aerobic or anaerobic at 20 °C for up to six months.

Figure 3.1 *Microbial challenge of test strips*

3.3.3 Environmental challenges

Appendix A3 gives the environments to which both sealant strips and joints were exposed. These ranged from aerobic/anaerobic sewage to raw and treated water sites.

Test strips of each sealant formulation (three for each time-point) with the appropriate controls were threaded onto nylon and suspended in a wire basket subsequently sealed with plastic mesh (Figure 3.3). Each basket was immersed in 3 ppm hypochlorite for 10 minutes before placement in the challenge environment.

Indelibly marked joints were also placed in wire baskets, treated with hypochlorite, and immersed in the challenge environment.

3.3.4 Microscopy

On removal from the test regime sealant samples (strips) were gently rinsed with tap water before photography using light and scanning electron microscopy.

3.4 EVALUATION OF RESULTS

3.4.1 Results of strip tests

Before trying to interpret the results in terms of the effects of the principal variables, it is necessary to have some measure of the test errors involved. The test errors can arise from many causes, including impure raw materials, mixing inefficiency, inaccuracies in the strip dimensions, the presence of surface flaws and other crack initiating sites, voids within the test piece, test temperature, inconsistent clamping and variable slipping in the testing machine or calibration errors.

On averaging the results of a triplicate strip test set (Figure A1.1), an estimated standard deviation was calculated. This quantity would normally be considered of little significance for such a small sample size, but it was found that this standard deviation fell within quite a narrow range (±5–6 per cent of the mean figure) over many of the test strip triplicate results. This suggested that the overall errors inherent in the test itself could be reasonably predicted from this figure. Since, for a normal distribution, there is a probability of 0.95 that a value lies within two standard deviations of the mean, it was decided that an error of no more than ±10 per cent of the mean value should be expected. This range was calculated for every strip test triplicate result at the six-month time period.

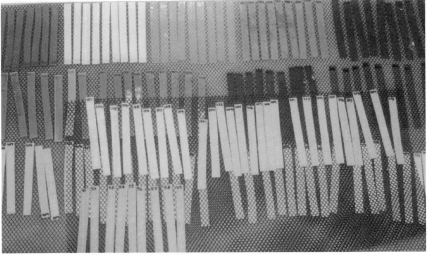

Figure 3.2 *Assembly of test strips for environmental challenge*

Clearly, a comprehensive test for microbiological degradation of a strip would show a gradual change with time, and results should be taken at several time-points for this purpose. For this study, this was not possible, as it would have magnified the number of tests beyond the capability of the programme. Only two time-points were taken (three and six months), to see if consistent changes could be detected within this period. Thus, the first time-point value ranges (within ±10 per cent of the means) were compared with the second time-point value ranges (within ±10 per cent of the means) (Figure 3.3).

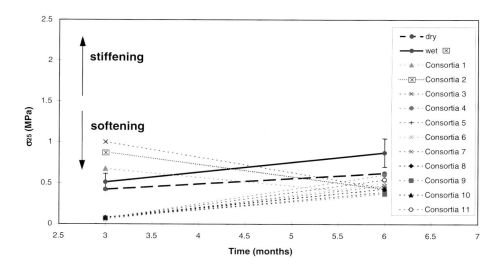

Figure 3.3 *Relationship between σ_{25} and the exposure time of PU2 to different microbial consortia*

If the first time-point values were less than the second time-point values (ie with no overlapping of the ranges of these results), this was termed a positive change. Conversely, if the first time-point values were greater than the second (with no range overlap), this was termed a negative change. If there was overlapping of the two ranges, there was deemed to be no significant difference between the values obtained.

Of more significance in detecting a change as a consequence of microbiological attack, the first and second time-point values should be compared with the corresponding values for the wet controls. A positive change for modulus here would indicate a hardening of the sealant from microbial attack, whereas a negative change would indicate a softening. Typical changes are illustrated diagrammatically in Figure 3.4.

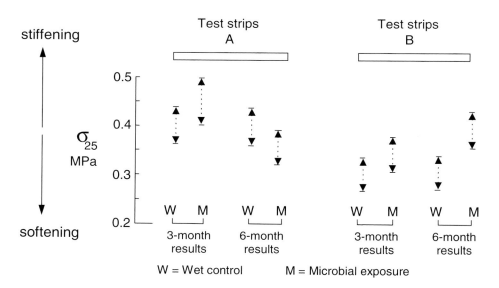

Figure 3.4 *Method of evaluating strip test results for σ_{25}*

In Figure 3.4, Test strips A would be considered to show insignificant change from the wet controls at both time-points. However, whereas the wet control had shown no significant change between three and six months, the exposed sample had shown a significant decline in modulus (ie a negative change).

Test strips B were not significantly different from the wet controls at the first time-point, but showed a significantly increased modulus (a positive change) at the second time-point. Neither the wet controls nor the microbially exposed sample showed significant change between the three- and six-month time-points.

Clearly, further time-points would have made any trend in results more apparent.

In the case of sealant PU2 illustrated in Figure 3.3, consortia 1, 2, and 3 have had a significant influence on this sealant.

These assessments of σ_{25} changes have been summarised in tabular form for all strip tests in Section 4.

3.4.2 Results of joint tests

The joint test results for ΔF_{25} and ΔE_{max}, summarised in Appendix A5, were further interpreted using the method described in the Standard WIS-4-60-01, as described in Appendix A1.

3.4.3 Results of microscopic (SEM) examination of sealant surfaces at three and six months

The techniques and procedures that have been used to ascertain/evaluate the extent of biofilm formation and the physiological activity of the component species are many and varied. However, few give any indication of the impact of the biofilm on the substratum *per se*. In this programme the impact of various consortia on selected polymeric sealant formulations was assessed by mechanical measurement of modulus and strength changes over time, together with visual examination of the sealant surface by microscopy. No other procedure for the assessment of microbial activity in biofilms was deemed to be of value in ascertaining the degradative potential of consortia or environmental challenges.

Changes in surface morphology, as ascertained by scanning electron microscopy, were categorised as follows (Figure 3.5; Table 7.5):

- No change   no perceived change when compared to the wet control
- Holes   holes apparent on the sealant surface
- Spongy   surface has an overall open, sponge-like, appearance
- Cracks   cracks, of varying depths and extent apparent on the sealant surface
- Crumbly   manifestation of extensive cracking of the surface.

Further details are given in Appendix A1.

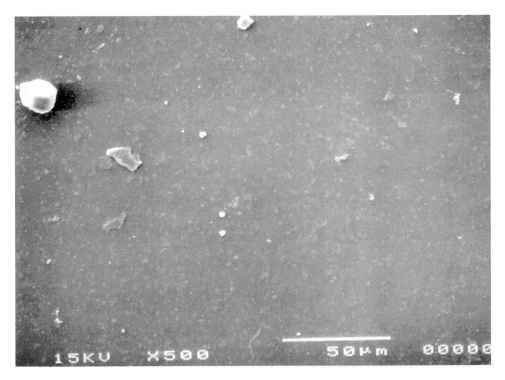

(a) Polyurethane 1: wet control (six months), magnification × 500

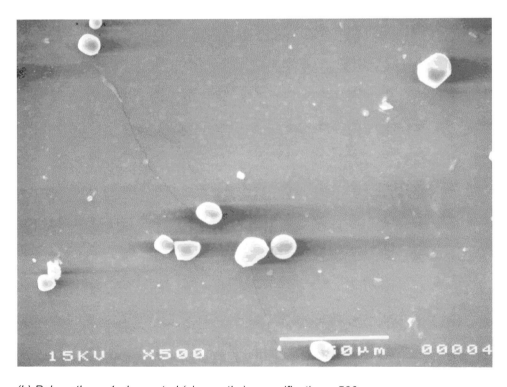

(b) Polyurethane 1: dry control (six months), magnification × 500

Figure 3.5 *Changes in surface morphology of sealant strips, as ascertained by scanning electron exposed to different microbial consortia and environmental challenges*

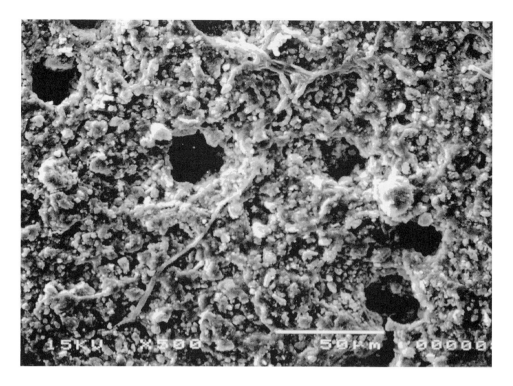

(c) Hole formation: Polyurethane 1 challenged with Consortium 2 (three months), magnification × 500

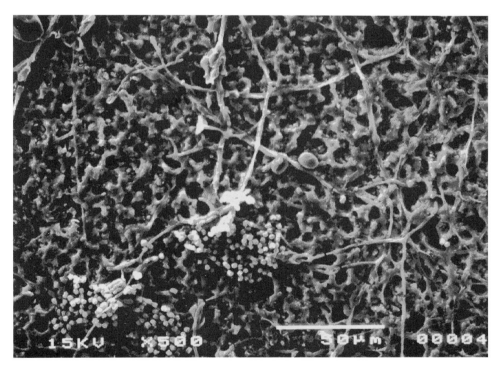

(d) Hole formation: Polyurethane 1 challenged with Consortium 2 (six months), magnification × 500

Figure 3.5 *Changes in surface morphology of sealant strips, as ascertained by scanning electron exposed to different microbial consortia and environmental challenges (continued)*

(e) Crack formation leading to crumbling: Polyurethane 1; Tewkesbury pre-treatment (three months), magnification × 75

(f) Crack formation leading to crumbling: Polyurethane 1; Tewkesbury pre-treatment (six months), magnification × 500

Figure 3.5 *Changes in surface morphology of sealant strips, as ascertained by scanning electron exposed to different microbial consortia and environmental challenges (continued)*

(g) Spongy effect: Polyurethane 1; 99 per cent humidity/sewage (six months), magnification × 500

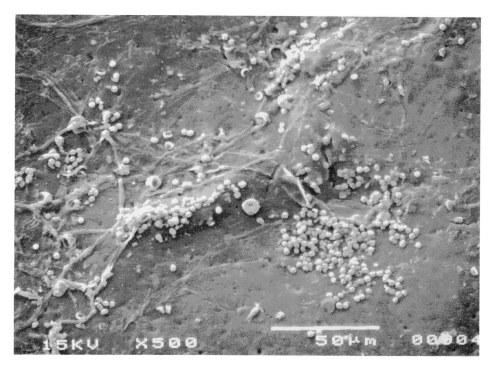

(h) Spongy effect: Polyurethane 1; consortium 1 challenged in soil (six months), magnification × 500

Figure 3.5 *Changes in surface morphology of sealant strips, as ascertained by scanning electron exposed to different microbial consortia and environmental challenges (continued)*

4 Experimental data and interpretation

4.1 REPORTED INCIDENTS OF MICROBIAL DEGRADATION

Visits were made to several sites at which there were suspected incidents of microbial degradation. Of these the following were significant.

4.1.1 Emley Moor and Foxroyd service reservoirs: Yorkshire Water

These potable water service reservoirs were approximately 18 months old. They are of concrete slab construction, the slabs being joined by a polymeric sealant. There was found to be substantial growth of a copious polysaccharide-producing micro-organism restricted, very noticeably, to the surface of the sealant, ie it was not evident on the concrete proximal to the sealant. There was no biofilm on the sealant above the water interface.

Close visual examination of the polymer did not show any signs of degradation, implying that the polymer surface was being exploited simply as a substratum for biofilm formation. However, with time, compromising of the polymer may become apparent. The causative organisms of the polysaccharide secretion were *Pseudomonas fluorescens* and *Pseudomonas putida*. These were subsequently used in the microbial consortia challenge experiments.

From the age of these reservoirs it is probable that the sealant biofilm was utilising substrates leaching from the sealant, which would explain its restriction to the sealant surface. The copious amounts of polysaccharide indicate a high carbon to nitrogen ratio; ie the population was nutrient-limited with respect to a fixed nitrogen source.

4.1.2 Sunnyside and Diamond Avenue storage reservoirs: Severn Trent Water

On routine cleaning of these reservoirs, which are more than 20 years old, it was observed that the lead dioxide polysulfide sealant had been compromised.

The sealant was found to be very powdery (Figure 4.1) to a depth of 1–2 mm. At greater depths it did not appear to be compromised. Only immersed sealant was affected. As in the case of the Yorkshire storage reservoirs a *Pseudomonas* species was isolated from this material. Chemical analysis of the compromised sealant is summarised in Table 4.1.

Table 4.1 *Analysis of sealant from Sunnyside and Diamond Avenue reservoirs compared to a typical lead dioxide-cured polysulfide sealant*

Analysis	Sample RS2 per cent by weight	Sample RS3 per cent by weight	Typical lead dioxide-cured sealant, per cent by weight
Calcium (Severn Trent Laboratories)	2.7	2.1	
Calcium carbonate (calculated)	6.75	5.25	34
Lead (Severn Trent Laboratories)	2.9	2.4	
Lead dioxide (calculated)	3.3	2.77	2.7
Polymer content (Morton International)	4.51	3.21	36
Plasticiser (Morton International)	19.77	24.14	23
Ash content (Morton International)	69.43	68.14	40

Figure 4.1 *Compromised sealant at Sunnyside Storage Reservoir (Severn Trent Water). The surface of the sealant was very powdery to a depth of 1–2 mm. This surface layer was easy disturbed and removed. The underlying sealant appeared to be unaffected*

From these data it is apparent that both the filler (calcium carbonate) and polymer content have been significantly reduced, presumably by microbial degradation. The data in Table 4.1 are chemical analyses and the sum of the component analyses will not be 100 per cent.

4.1.3 Yearby Service Reservoir: Northumbrian Water

This reservoir was constructed in 1972, the joints being sealed with caulkite polysulfide. At an inspection in the early 1980s it was discovered that the floor (immersed) joints had suffered severe degradation. At the time of the inspection there was an overlying layer of silt. When the silt was removed, the exposed surface of the polymer was found to be very soft, and there was a strong sulfide smell. In this situation, the sealant has almost certainly been exposed to anaerobic conditions, the sulfide being derived, therefore, from the action of *Desulfovibrio* species.

The reservoir was cleaned but no action was taken with respect to the sealant. Access for re-examination/sampling was not possible during the tenure of this programme.

4.1.4 Tilbury Sewage Treatment Works: Anglian Water Services

Extensive demolition and refurbishment of sludge tanks was being undertaken at this site. These units were approximately 12 years old.

Several samples of the polymeric sealants used were taken for microbiological examination. Unfortunately, the sealants had been allowed to dry out for some time before sampling and consequently little of microbiological value was forthcoming. On visual inspection there was little evidence of polymer breakdown other than where it had become detached from the enamel as a consequence of rusting at the joint.

4.2 RESULTS OBTAINED WITH POLYURETHANE SEALANT STRIPS

4.2.1 Dry and wet controls

The summary results (Appendix A4) for modulus at the two time-points show two unexpected results. (i) In all cases the wet and dry controls have increased in modulus between the three-month and six-month time-points; (ii) the wet control is stiffer than the dry control in most cases (the exceptions in five of the 18 cases being PU2(b) (aliphatic), PU3 (standard), PU6(b) (phthalate plasticiser), PU8(a) (polymer only – polyether), PU9 (polymer only – polyester)).

The most obvious explanation for (i) is that the curing reaction is still proceeding between these time-points, and the most reasonable explanation for (ii) is that curing in the wet is faster than curing in the dry, at least in some cases.

These observations mean that the wet results are complicated by the fact that water absorption and curing are proceeding at the same time. Since none of the samples appears to show much Phase 2 water absorption (from weight and volume changes), most of the softening due to water absorption would have been observed during the first day or two of immersion, when Phase 1 water absorption occurred. Thus, there would be little difference in water content between three and six months, and the rise in modulus must therefore mainly reflect the continuation of curing.

Clearly, if the two-part cure of these materials were complete before exposure to water, there would be no further chemical effect arising from exposure to water. This has been illustrated diagrammatically in Figure 2.1. The later stages of the dry cure, however, would be very slow since it would depend on the diffusion of very large isocyanate-terminated molecules. If the cure were not complete when immersed in water, water would be competing with the polyol for the isocyanate, a contest that the water would probably win because of its greater mobility and molar concentration. This would result in a polymer containing urea links rather than urethane links, a change that would be expected to cause stiffening relative to the properly cured sealant. The stiffening could be severe if it resulted in phase separation in the product. The polyol that had not reacted fully would remain as a softener. These changes would result in modulus behaviour corresponding to some version of Figure 2.2.

Even the dry-cured sealants would be expected to suffer from this effect if the late stages of the curing reaction were unduly slow, since moisture would diffuse in and be competitive with the polyol for the available isocyanate.

4.2.2 Effect of fungal consortia (C1 and C2) and anaerobic bacterial consortium C3 on polyurethane sealants

The most striking effect seen from an examination of the summary results is the very high modulus seen at the three-month time-point. This occurs with all the polyurethane sealants with only two exceptions – the C1 values for PU6 (phthalate plasticiser) and PU7 (no clay). C2 and C3 consortia show the effect very strongly. The effect is so marked that the possibilities of errors in the sealant formulation and in the test methods were examined. It was decided that the sealant formulation could not have been in error because samples were randomised from the same batch as used for all the other consortia, and for the wet and dry controls. Moreover, these three consortia showed the same effect in virtually all of the polysulfide and silicone samples. The possibility of testing error in these samples was investigated in an examination of the broken test pieces, albeit over a year later when they had dried out. Although they were not tested by machine, the polyurethane samples examined were all noticeably stiffer than the wet controls when pulled manually. The original traces from the testing machine and calculations from them were also checked, but no error was found. One must conclude that there is a real and strong stiffening effect arising during the first three months of exposure to these consortia, which is substantially independent of the sealant composition.

It seems logical to suppose that this effect cannot be a chemical one, in view of the wide variety of compositions over which it occurs. Such stiffening, if it involved attack on sealant components, would need to be either a cross-linking effect or the removal of plasticiser from the sealant. Since some of the sealants affected do not contain plasticiser, this latter possibility is ruled out. No universal cross-linking mechanism can be envisaged.

It would appear, therefore, that the effect is a purely mechanical one, probably involving the anchorage of a biomass layer to the surface of the sealant in the early stages of attack. The composition of the sealant appears to be irrelevant to this attachment, although the question of nutrition available to the organism still remains.

Even if such a mechanism for the stiffening is accepted, an explanation for the subsequent softening is still required. This must involve a very rapid and strong degradation of the sealant, since the modulus at six months is generally below that of the wet and dry controls, and possibly still declining rapidly. The attrition process here is not immediately obvious, even from the microscopic examination where there seems to be no clear correlation of the C2 or C3 consortia with any particular surface features.

4.2.3 Effect of bacterial consortia C4–C11 on polyurethane sealants

In a sense, sealants exposed to these consortia show the opposite effects from those exposed to consortia C1–C3. They generally show abnormally low moduli at the three-month time-point, well below those of the wet and dry controls, but rise to more normal levels by the six-month time-point. Again the effect seems to be fairly universal for the polyurethane sealants, but this time it is not obviously also the case for the polysulfide and silicone sealants (discussed later). The low three-month modulus values cannot be explained by microbial degradation of the polymer, since the trend is reversed by the six-month point. A more reasonable explanation might be a retardation of the cure rate (compared with the wet and dry controls). This could possibly be due to the removal of catalyst by the micro-organism, possibly through a preferred chemical interaction. It is known that the organometallic catalysts, particularly the mercury ones, have anti-bacterial effects; they might, conceivably be removed/diluted by microbial activity.

4.2.4 Effect of environmental exposures on polyurethane sealants

Unlike the consortia results, the environmentally exposed samples do not generally exhibit the anomalous high three-month values. One or two notable exceptions, for which no obvious explanation can be offered, are: PU2 (aliphatic) under anaerobic sewage (*Sew anaer*) conditions, PU7 (no clay) in the sewage outflow (*Sew out*) and PU9 (polymer only – polyester) for *Yorks post*, *Soil* and *Sew 99% hum*.

In general, the environmentally exposed samples show a moderate increase in modulus between the three- and six-month time-points, generally of an order consistent with furtherance of the cure. However, it is very noticeable that the Tewkesbury post-treatment (*Tewkes post*) samples show consistently very low values of modulus, both at the three- and six-month time-points. The modulus levels are much below those of the wet and dry controls, suggesting a serious level of degradation. It cannot be due to initial undercure of the samples, since they were cut from the same cast and cured sheets as all the other samples. It must be that either degradation involving polymer chain scission occurs quite readily in this environment, or the samples have for some reason absorbed more water. The worst-affected sample appears to be PU9 (polymer only – polyester), so perhaps attack by hydrolysis of ester groups is significant. For this sample, both the Tewkesbury pre-treatment and post-treatment waters appear to be disastrous. It also appears to be suffering degradative effects, albeit less severe, from three other environments (sewage at 99 per cent humidity, soil and Yorkshire post-treatment).

Several of the sealants (PU2 (aliphatic), PU3 (standard), PU5 (mercury catalyst), PU6 (phthalate plasticiser) and PU8 (polymer only – polyether)) show that most of the environmental exposures cause consistent stiffening compared with the wet control. However, there is no clear correlation with any of the structural variables (except to note that PU1 (polyester) shows moduli generally lower than the wet control, again suggesting possible increase in rate of hydrolysis of the ester groups by microbial attack). The stiffening that occurs with most samples could be due to mechanical effects from stiff microbial growths or residues.

The removal of clay from PU3, as has been done in PU7 (no clay) appears to change the modulus behaviour very considerably. Particularly bad are the *Yorks post* and *Tewkes pre* figures at three-months, where modulus is so low as to be unmeasurable. Clearly, the presence of filler has important effects that have not yet been recognised.

4.3 RESULTS OBTAINED WITH POLYURETHANE SEALANT JOINTS

The joints made using polyurethane sealants were as follows:

- PU3 (standard) sealant, epoxy primer
- PU3 (standard) sealant, isocyanate primer
- PU6 (phthalate plasticiser), epoxy primer

In the following sections the test results of these will be compared first between themselves and second against the results for strip tests in corresponding environments.

The main criteria generally used for assessment of the performance of sealant joints after exposure to wet conditions are the change in force (or stress) at 25 per cent extension and the change in extension at maximum force (or stress), both relative to the dry control. In the present case, however, we are concerned with the relative performance of the environmentally exposed samples against the wet control, since we are looking for indications of microbial degradation rather than softening by water. The main changes to be examined are therefore the modulus change $\Delta\sigma 25$ and the extensibility change ΔE_{max}, relative to the wet control. These have been calculated for all the PU3 tests, but only in the extension tests at Time-point 2 for the PU6 tests, where some wet control figures are unavailable. The figures are tabulated in Table 4.2, below.

Table 4.2 Changes in (a) 25 per cent modulus ($\Delta\sigma_{25}$) and (b) extension at maximum force (AE_{max}), relative to the wet controls, for polyurethane sealant joints exposed to various environments. (N = no specimens received or specimens not identified, F = premature failure of joint)

(a) $\Delta\sigma_{25}$ relative to wet control (per cent)

	Time point	Dry	Soil	Yorks pre-	Yorks post-	Tewks pre-	Tewks post-	Sew 99%	Sew out	Sew anaer
PU3/epoxy	1	+3.2	-2.4	-39.0	+12.9	-17.8	-2.4	-20.5	-14.6	N
	2	+23.2	-5.0	+7.5	-8.3	-1.3	-4.3	F	-12.5	-11.8
PU3/isocy	1	+14.4	-6.9	-4.8	+3.2	-9.9	+1.9	-12.9	-13.4	N
	2	+9.7	F	N	+7.1	N	N	F	F	F

(b) AE_{max} relative to wet control (per cent)

	Time point	Dry	Soil	Yorks pre-	Yorks post-	Tewks pre-	Tewks post-	Sew 99%	Sew out	Sew anaer
PU3/epoxy	1	-20.4	+3.7	-14.8	-7.4	-27.8	-18.5	-35.2	-33.3	N
	2	0	-22.2	-13.3	+8.9	-11.1	-17.8	-51.1	-53.3	-15.6
PU3/isocy	1	-15.7	-29.4	-47.1	-47.1	-47.1	-45.1	-52.9	-49.0	N
	2	+5.7	-42.9	N	-2.9	N	N	-48.6	-37.1	-40.0
PU6/epoxy	2	N	-27.8	+38.9	+11.1	-16.7	-5.6	-38.9	-38.9	+33.3

4.3.1 Comparative performance after different environmental exposures

The following comments are made after reference to Table 4.2 for the first and second time-points.

It can be seen that the wet control moduli for PU3 are always lower than those of the dry controls. This is in contrast to those obtained from the sheet samples, where most of the wet controls showed higher stiffness at both time-points. Although there is a lack of consistency in the order of the change between the two time-points, the results suggest that the joints were more fully cured before testing and the loss in modulus compared with the dry controls is due mainly to water absorption. Unfortunately, there are no corresponding results available for PU6. The extensibility changes between wet and dry controls are very erratic, probably due to the rather brittle nature of the sealants.

Table 4.2 also shows that, with a few exceptions, the modulus of the PU3 sealant indicates a negative change with environmental exposure, suggesting a degradative attack. There are indications in some cases that the initial softening may be followed by a hardening before failure occurs. There are more failures of isocyanate-primed joints (Time-point 2) than of epoxy-primed ones. This may be due to chemical and/or microbial degradation of the primer in the former case.

The poorest performance of the sealants is seen in the *Sewage* environments, where large reductions in both moduli and extensibility occur. The *Soil*-exposed samples also show a progressive deterioration on all cases. It is more difficult to assess the performance of the *Yorks* and *Tewkes* samples, where more results are missing and the large extension losses at the early time-point are probably due to failure of the primer.

4.3.2 Comparison of strips and joints after different environmental exposures

It has already been noted that the differences between wet and dry controls suggest that the joints may be more fully cured at the start of the tests than the strip specimens. This may well be due to the unavoidable delay between making and testing the joints, which therefore had more time to cure. Some of the modulus change differences seen when comparing strips and joints may be partly due to this effect.

Since there were no primers present in the strip tests, the only real comparison available is of PU3 (standard) in both strip and joint (PU3/epoxy) form (the PU3/isocyanates are probably suspect for comparison because of the primer weakness). This comparison may best be seen in the form of Table 4.3.

Table 4.3 *Comparison of (a) $\Delta\sigma_{25}$ and (b) ΔE_{max} results for PU3 in strip and joint form*

(a) $\Delta\sigma_{25}$ relative to wet control (per cent)

	Time-point	Dry	Soil	Yorks pre-	Yorks post-	Tewks pre-	Tewks post-	Sew 99%	Sew out	Sew anaer
PU3 strips	1	+20.7	+65.5	+48.3	+34.5	+44.8	-27.6	+72.4	+48.3	+51.7
	2	+17.9	+12.5	+21.4	+8.9	+17.9	-55.4	+28.6	-21.4	+32.1
PU3/epoxy joints	1	+3.2	-2.4	-39.0	+12.9	-17.8	-2.4	-20.5	-14.6	N
	2	+23.2	-5.0	+7.5	-8.3	-1.3	-4.3	F	-12.5	-11.8

(b) ΔE_{max} relative to wet control (per cent)

	Time-point	Dry	Soil	Yorks pre-	Yorks post-	Tewks pre-	Tewks post-	Sew 99%	Sew out	Sew anaer
PU3 strips	1	-372	-14.1	-36.2	-56.2	-53.7	-48.2	-48.8	-50.7	-37.9
	2	+47.4	+27.6	+33.9	+52.9	-16.6	+68.1	-25.5	-15.9	-49.3
PU3/epoxy joints	1	-20.4	+3.7	-14.8	-7.4	-27.8	-18.5	-35.2	-33.3	N
	2	0	-22.2	-13.3	+8.9	-11.1	-17.8	-51.1	-53.3	-15.6

Comparison of the values for strips and joints in the above figure shows a rather disappointing level of correlation. Whereas the modulus results for both time-points show mainly a hardening in the strip tests, they show mainly softening in the joint tests. The extensibility tests vary widely in the strip tests and their absolute values clearly cannot be relied upon. This is probably due to the sensitive nature of this sealant to brittle fracture.

It can be said that, in general, the soil-exposed samples are probably the best performers. Those in the three sewage environments are consistently poor, showing fairly large departures from the wet control values.

4.4 RESULTS OBTAINED WITH POLYSULFIDE STRIPS

4.4.1 Dry and wet controls

The summarised results for all of the polysulfide strips show in this case the expected results that the dry controls have a consistently higher modulus than the wet controls. There is a slight positive change between the three- and six-month time-points in almost every case, indicating that some additional curing is still progressing.

It is gratifying to note that the results for PS1 (manganese-cure, chlorinated plasticiser) controls are virtually the same as those for PS4 (manganese cure, chlorinated plasticiser, thixotropic), as would be expected in view of the very slight difference in formulation. Comparing the results for PSI (manganese-cure, chlorinated plasticiser) and PS2 (lead-cure, chlorinated plasticiser) shows that dry modulus of the lead-cured material is substantially higher than the manganese-cured, whereas there is little difference between the wet control moduli. This could indicate either that the lead-cured samples are at a higher state of cure or that the lead-cured compounds are inherently stiffer for some other reason (eg different ratios of different cross-link types – see Section 2.3.1). The former seems the more likely, as it is well known that cure rate with lead dioxide is high and markedly activated by the presence of water.

Results for PS2 (lead-cure, chlorinated plasticiser) and PS3 (lead-cure, phthalate plasticiser) provide a comparison of chlorinated *versus* phthalate plasticisers. The superior action of phthalate as a plasticiser is seen in the lower modulus of the dry controls in PS3. In the wet state, this difference virtually disappears, presumably because of the much more powerful plasticising action of water.

In the case of PS5 (simple polysulfide) it is noticeable that the proportionate difference between dry and wet is lower at both time-points. This is almost certainly due to the reduced level of water absorption in the presence of only a very small amount of filler.

4.4.2 Effect of fungal (C1 and C2) and bacterial (C3) consortia on polysulfide sealants

As was the case with the polyurethanes, consortia C1 and C2 always appear to produce an anomalously high value for modulus at the three-month time-point. Consortium C3, which also gave universally high values in the polyurethane, now gives high values only in the lead-cured (PS2 and PS3) and simple polysulfide (PS5) sealants. The two remaining sealants PSI and PS4, both manganese-cured, do not show this high modulus value with consortium C3. It might be concluded from this that there is something unique in these formulations that inhibits fungal growth on the surface. Clearly, it must be something to do with the cure system, a likely possibility being that manganese or permanganate ions act to destroy the fungus in this particular case.

If this is indeed the explanation, it would seem that the micro-organisms present in consortia C1 and C2 are not affected by these ions, since they both show large increases in modulus.

Although PS2, PS3 and PS5 all show anomalous high values with all three consortia C1–C3, the modulus values are not all in the same order. In the case of PS5 (simple polysulfide) a fourth consortium (C10) also shows the effect. Possibly both lead and manganese can suppress the growth of a surface biomass of this consortium.

4.4.3 Effect of bacterial consortia C4–C11 on polysulfide sealants

In almost all cases, there is a positive change in modulus between the three- and six-month time-points. By the six-month time-point, almost all of these consortia have produced moduli that are higher than that of the wet control, and approaching the dry control value in some cases. Although this must be a significant observation, there does not seem to be a clear-cut relation between the magnitude of the change and the consortium type. It might be said as a generalisation, however, that some consortia, such as C10 and C8, produce a greater change than others, such as C5 and C6.

It seems that the bacterial consortia are in general exerting some action that serves to increase the modulus of the wet-cured sealants. The effect could be either one of accelerating the cure, which seems unlikely as the sealants must be very well cured by the six-month time-point, or one of degradation and removal of plasticiser. The alternative of cross-linking the polymer is considered unlikely, as no possible mechanism can be envisaged. Removal of plasticiser from the surface regions would be expected to harden and shrink them, possibly producing the type of craze pattern which has been observed by microscopy.

Surprisingly, there is no real evidence that any of these bacteria can attack the polymer backbone, at either the formal or disulfide groups. The resulting chain scission would be expected to produce a sharp reduction in modulus, for reasons discussed earlier.

4.4.4 Effect of environmental exposures on polysulfide sealants

It can be seen from the summary results that most of the modulus values for the first time-point are greater than those for the wet control. However, it must be emphasised that the wet controls were exposed under laboratory conditions, and are not, therefore, strictly comparable. It may be that the environmental samples have suffered less swelling (and therefore retain higher modulus) because of effects of different pH or salt content of the water. The values are generally low and therefore subject to some inaccuracy. It cannot be said for certain, therefore, that these results generally indicate a stiffening from microbial attack.

There are some notable exceptions, however, particularly in the case of PS1 (*Sew out* and *Sew anaer*). Here the values are much higher than either the wet or the dry controls. Moreover, the E_{max} values are much lower than the control values. These show a pronounced stiffening in these cases. However, the almost identical formulation PS4 does not show this stiffening effect to a marked degree. No immediate explanation for this can be seen.

There are several other cases where the first time-point modulus values are anomalously high, even in comparison with the dry controls. The most consistent of these is the value for *Soil* exposure, where in all cases a higher than dry modulus is seen. For these, the E_{max} values are generally lower than those of the dry controls, and it is clear that a stiffening process is occurring. The most likely explanation would again appear to be a fungal biomass forming on the surface.

At the second time-point, the most consistently stiffened sample is again the soil-exposed one, where modulus is higher than that of the dry control in each case. There seems to be no positive correlation between the degree of stiffening and the type of formulation, indicating again a more mechanical effect.

With a few exceptions, the second time-point moduli for PS1 and PS4 (both manganese-cured) are generally between those of the wet and dry controls. These, together with the retention of reasonably high E_{max} values, indicate little microbiological activity. For the lead-cured samples PS2 and PS3, however, there are several cases where complete failure of the specimens has occurred. Where it is measurable, the modulus is erratic and all of the measured extension values are lower than those of the wet controls. Thus, the lead-cured samples appear much more susceptible to attack in these environmental situations. Furthermore, the proportion of failures is much higher in PS3 (lead-cure, phthalate plasticiser) than in PS2 (lead-cure, chlorinated plasticiser). All of these observations are consistent with the expected order of results discussed in Section 2.3.1. The results would suggest a surface attack, mainly at the plasticiser, which hardens the surface and at the same time generates stress raisers that allow fracture to occur more easily. Clearly, manganese provides more resistance to this than lead, and chlorinated plasticiser more than phthalate plasticiser.

For PS5 (simple polysulfide), however, there are five environments where the modulus is higher than the dry control and three where it is lower. Although all of the E_{max} values for PS5 are lower than that of the wet control, they are nevertheless reasonable values and show only little degradation. Possibly some surface hardening and crazing from bacterial attack is indicated, particularly in the *Sew 99%* and *Yorks post* cases, but effects are not severe, presumably partly because of the absence of plasticiser. The *Tewkesbury pre-* and *post-* environments do not show particularly severe attack on this sample as they did on the polyurethane PU9 (polymer only – polyester).

4.5 RESULTS OBTAINED WITH POLYSULFIDE SEALANT JOINTS

The joints prepared for testing using polysulfide sealants were as follows:

- PS1 (manganese cure, chlorinated plasticiser) sealant, epoxy primer
- PS1 (manganese cure, chlorinated plasticiser) sealant, isocyanate primer
- PS3 (lead cure, phthalate plasticiser) sealant, epoxy primer

These were selected to facilitate a comparison between the best and worst cases for microbial attack, and to compare epoxy against isocyanate primer.

As with the polyurethanes (Section 4.3), the main criteria used to assess their relative performance were the values of changes in 25 per cent modulus ($\Delta\sigma_{25}$) and maximum extension (ΔE_{max}), relative to the wet controls. These values have been calculated from the detailed results and are presented in Table 4.4, below. Values cannot, of course, be tabulated for those cases where no wet control results were available.

Examination of the $\Delta\sigma_{25}$ figures for PS1 shows that the dry controls are again stiffer than the wet, as would be expected from a fully cured sealant that absorbed water after immersion. As expected, the difference is greater at the second time-point when more water has been absorbed. This stiffness in the dry state would have been expected to result in a reduced value of ΔE_{max}, but this is clearly not the case. In fact, some very large values of ΔE_{max} were observed in PS1/epoxy, and even larger values for PS1/isocyanate. The reason for this can be seen in the erratic nature of the wet control failures. In the case of PS1/epoxy these have unfortunately given two out of three adhesive failures at low extensions at the first time-point, and a similar proportion at very low extensions at the second time-point. PS1/isocy gives even poorer wet control results with interfacial failure, especially at the second time-point This type of interfacial failure was much less obvious in the dry controls (except PS1/isocy at the second time-point), hence the abnormally large differences between the wet and dry samples. Some abnormal values are also seen in the environmentally exposed samples.

These are believed to be due to the same effect, ie the erratic nature of the changes between cohesive- and adhesive-type failures.

Table 4.4 *Changes in (a) 25 per cent modulus ($\Delta\sigma_{25}$) and (b) maximum extension (ΔE_{max}), relative to the wet controls, for polysulfide sealant joints exposed to various environmental conditions. (N = no specimens received or specimens not identified, F = joint failed at <25 per cent extension)*

(a) $\Delta\sigma_{25}$ relative to wet control (per cent)

	Time point	Dry	Soil	Yorks pre-	Yorks post-	Tewks pre-	Tewks post-	Sew 99%	Sew out	Sew anaer
PS1/epoxy	1	+3.7	-1.2	-5.1	+0.5	+0.1	-1.2	-2.7	-2.0	N
	2	+29.0	+1.3	-5.1	0	N	+6.3	-5.1	0	-11.4
PS3/epoxy	2	N	+14.5	+242.1	+7.9	-15.8	-21.1	-97.4	-28.9	-39.5

(b) ΔE_{max} relative to wet control (per cent)

	Time point	Dry	Soil	Yorks pre-	Yorks post-	Tewks pre-	Tewks post-	Sew 99%	Sew out	Sew anaer
PS1/epoxy	1	+28.5	+37.5	+10.2	+27.3	-9.7	-4.4	-9.5	-12.4	N
	2	+294.8	+240.7	+144.8	+176.2	N	+129.7	+25.0	+108.1	+29.7
PS1/isocy	1	+2386	+871.4	+114.3	+200.0	-9.5	+109.5	-47.6	+295.2	N
	2	+3000	+1943	+1429	+728.6	+14.3	+414.3	F	+971.4	+157.1
PS3/epoxy	2	N	+5.7	-52.8	+5.7	+67.9	+41.5	-26.4	-22.6	-18.9

4.5.1 Comparative performance after different environmental exposures

Despite the above remarks, it is still possible to compare the relative performance of the differently exposed samples from Table 4.4. The $\Delta\sigma_{25}$ values for PS1/epoxy are probably the most stable set of results available. At the first time-point they vary within about ±5 per cent of the control value; at the second time-point within a rather broader range. The *Sew 99%t* and *Sew anaer* results would probably be considered the poorest in this set, but it is difficult in view of the nature of the results to arrange the remaining samples in any logical order.

Modulus changes are not shown in the table for the PS1/isocy joints, due to the failure of both wet controls below the 25 per cent limit. Examination of the detailed results shows that such failure also occurs in two other cases (*Tewkes pre* and *Sew 99%*) at the first time-point, and in three other cases (*Tewkes pre*, *Sew 99%* and *Sew anaer*) at the second time-point. All other modulus results are similar in magnitude to those of the corresponding PS1/epoxy joints. This is, of course, as would be expected since the sealant is the same in both cases. Thus, the differences between the PS1/epoxy and PS1/isocy are entirely due to the primers, of which the isocyanate is the much more suspect (except in the dry state where they perform comparably).

The PS3/epoxy results at the first time-point are not included because of the absence of wet control test results. However, examination of the F_{25} values from the detailed results shows a force decrease in the order:

Yorks pre > Soil > Yorks post > Tewkes pre = Tewkes Post >> Sew out > Sew 99%

It is gratifying to note from Table 4.4 that virtually the same order is seen at the second time-point.

It seems that the most severe environments are the same for both sealant PS 1 and PS3, despite their chemical differences (different curatives and plasticisers). It seems that all three sewage environments are destructive in a sense that reduces modulus, indicating that attack upon the polymer is quite likely. The *Sew 99%* and *Sew anaer* are generally more corrosive than *Sew out*. The other environment that gives noticeably inferior results is *Tewkes pre*, but this is not generally as bad as the sewage environments.

4.5.2 Comparison of strips and joints after different environmental exposures

If PS1/epoxy results may be taken as typical of the polysulfide sealant joint behaviour, we can compare these with the results of strip tests in the same environments by reference to Table 4.5, below. In view of the very low and imprecise values of modulus for the strips, these values are shown directly in the table.

Table 4.5 Comparison of (a) σ_{25} or $\Delta\sigma_{25}$ and (b) AE_{max} results for PS1 in strip and joint form. (σ_{25} (wet) = 0.00 for time-point 1 and 0.02 for time-point 2). (N = No specimens received or specimens not identified)

(a) σ_{25} (strips) (NM-2), or $\Delta\sigma_{25}$ (joints) relative to wet control (per cent)

	Time point	Dry	Soil	Yorks pre-	Yorks post-	Tewks pre-	Tewks post-	Sew 99%	Sew out	Sew anaer
PS1 strips	1	0.05	0.11	0.04	0.04	0.04	0.02	0.03	0.30	0.34
	2	0.08	0.08	0.05	0.04	0.02	0.01	0.07	0.01	0.03
PS1/epoxy joints	1	+3.7	-1.2	-5.1	+0.5	+0.1	-1.2	-2.7	-2.0	N
	2	+29.0	+1.3	-5.1	0	N	+6.3	-5.1	0	-11.4

(b) ΔE_{max} relative to wet control (per cent)

	Time point	Dry	Soil	Yorks pre-	Yorks post-	Tewks pre-	Tewks post-	Sew 99%	Sew out	Sew anaer
PS1 strips	1	+130.1	+94.8	-2.4	+28.7	-34.6	+80.7	-82.8	-79.8	-81.8
	2	+165.4	+148.0	+4.1	+49.1	0.0	-47.9	-43.7	+20.4	-37.8
PS1/epoxy joints	1	+28.5	+37.5	+10.2	+27.3	-9.7	-4.4	-9.5	-12.4	N
	2	+294.8	+240.7	+144.8	+176.2	N	+129.7	+25.0	+108.1	+29.7

Again, the modulus results give the more reliable comparison, since they are (presumably) unaffected by the presence of primer in the joints.

If the modulus (strips) and modulus change (joints) results are ranked in decreasing order, the following comparison is seen:

First time-point

Strips: *Sew anaer > Sew out > Soil > Dry > Yorks pre = Yorks post = Tewkes pre > Tewkes post*

Joints: *Dry > Yorks post > Tewkes pre > Soil = Tewkes post > Sew out > Sew 99% > Yorks pre*

2nd time-point

Strips: *Dry = Soil > Sew 99% > Yorks pre > Yorks post > Sew anaer > Tewkes pre > Tewkes post = Sew out*

Joints: *Dry >> Tewkes post > Soil > Yorks post = Sew out > Yorks pre = Sew 99% > Sew anaer*

Clearly, there is no strong correlation between these rankings generally. If the second time-point is taken as a more realistic indication of degradation, there is a rather vague trend in which *Dry* and *Soil* are best and the various *Sewage* samples are poorest It is felt that many more samples would be required for a satisfactory correlation to be established.

The ΔE_{max} results give the following rankings:

First time-point

Strips: *Dry > Soil > Tewkes post > Yorks post > Yorks pre > Sew out > Sew anaer > Sew 99%*

Joints: *Soil > Dry > Yorks post > Yorks pre > Tewkes post > Sew 99% > Tewkes pre > Sew out*

Second time-point

Strips: *Dry > Soil > Yorks post > Sew out > Yorks pre > Tewkes pre > Sew anaer > Sew 99% > Tewkes post*

Joints: *Dry > Soil > Yorks post > Yorks pre > Tewkes post > Sew out > Sew anaer > Sew 99%*

Surprisingly, these results show a rather better correlation between strips and joints than the modulus figures. Again, the *Dry* and *Soil* samples are clearly least affected, the *Yorks* and *Tewkes* samples are intermediate, and the various *Sewage* samples are most affected. Exceptions to this trend are seen in the poor performance of *Tewkes post* and the better-than-expected results for the *Sew out*, both at the first time-point.

Considering both modulus and extensibility results, there is a clear general correlation between the strip and joint test rankings, particularly at the second time-points.

4.6 RESULTS OBTAINED WITH SILICONE SEALANT STRIPS

4.6.1 Wet and dry controls

Examination of the summary results shows that, except for Si1 (oxime-cured) dry control, there is evidence of further curing between the first and the second time-points. This is seen in a positive change in modulus and a universal reduction in E_{max} values. This again must be due to the fact that they are moisture-curing sealants, and highly dependent on humidity levels and temperature. In this case, they were shipped to us by the manufacturer, already enclosed in protective polyethylene sheets, which were not removed except when immersed. This might have retarded the dry cure to some extent.

It can also be seen that, at the six-month time-point, the wet control modulus is slightly higher than the dry control modulus in all three cases. This probably reflects a slightly higher degree of cure in the wet cases. There is not much evidence of the softening in water that might have been expected, either in these controls or, indeed, in any of the samples immersed in the consortia media. It would appear, therefore, that the silicone samples are highly resistant to the absorption of water.

4.6.2 Effects of fungal (C1, C2) and bacterial (C3) consortia on silicone sealants

It is seen that, again, the silicone sealants exhibit some anomalously high modulus values at the three-month time-point. The pattern here appears to show some dependence on sealant composition. Si1 (oxime-cure, unfilled) shows very high values in consortia C2 and C3, with moderately high values in consortium C1. Si2 (oxime-cure, filled) shows the anomalously high value only for C1, with only moderately elevated values for C2 and C3. Si3 (benzamide-cure, filled) shows a very high value in consortium C1, but only slightly raised values in C2 and C3. Surprisingly, the E_{max} values do not vary by much throughout. It seems reasonable to suppose that the very high C1 values correlate with the presence of calcium carbonate filler in the sealant, and that very high C2 and C3 values correlate with its absence. The calcium carbonate might be acting to provide calcium ions, regulate the pH, or to provide a mechanical support. It does not seem possible, however, to distinguish between these possibilities.

At the six-month time-point, these elevated values of modulus have disappeared and modulus levels are now among the lowest, even below that of the wet control in some cases. This suggests that the initial build-up of biomass that caused the high modulus at three-months has further progressed into more severe degradation of the sealant.

4.6.3 Effects of bacterial consortia C4–C9 on silicone sealants

The values of modulus observed at the three-month time-point for these bacterial consortia are generally very low and erratic (because of measurement difficulties). Most of them are lower than the wet control values. The E_{max} values, conversely, are generally very high. This combination is typical of undercure, although it is surprising that the polymers remain undercured after such a long period of immersion.

At the six-month time-point, modulus values are more realistic and there appear to be slight changes in modulus, compared with the wet control, which are fairly consistent C5 and C6 show rather small elevations of modulus, whereas C4, C7, C8, C9, C10 and C11 show more positive effects. The E_{max} values also reflect the same trend, E_{max} values generally being higher when the modulus is low and *vice versa*. These changes suggest a slight stiffening effect from the microbial attack, possibly a surface degradative effect such as removal of organic groups, which causes hardening.

There appear to be no clear-cut differences arising from the various silicone formulations.

4.6.4 Effects of environmental exposures on silicone sealants

At the first time-point, the modulus values are somewhat erratic but do not show very large differences. They are generally above the values for the wet controls, but these latter were measured in the laboratory and may not, therefore, be strictly comparable.

There is no clear evidence of undercure in these samples as there was in the case of the laboratory consortia tests. The *Tewkes pre* and *Yorks post* values seem distinctly lower than the others generally.

At the six-month time-point, the modulus values are in the main higher than those at three months, and the E_{max} values are slightly lower. This indicates a moderate hardening between the three- and six-month time-points, which may be due to further curing or to a microbial hardening. Exceptionally poor results are seen in the *Tewkes post* case for Si2 (oxime-cure, filled) and Si3 (benzamide-cure, filled), indicating a definite degradative effect here. Si1 (oxime-cure, unfilled) does not show this effect, which might therefore correlate with the presence of calcium carbonate filler. Conversely, the *Yorks post* value for Si1 (oxime-cure, unfilled) is somewhat low at six months, whereas the other (filled) silicones show a more normal response here. It is tempting to suggest that the same fungi may be present as in the consortia C1–C3, discussed above. Consistently good throughout are the results for *Sew 99%*, *Sew anaer* and *Soil*, all of which appear to show good extension values as well as high modulus, and do not show particular dependence on the sealant formulation.

4.7 RESULTS OBTAINED WITH SILICONE SEALANT STRIPS AND JOINTS

Only the sealant Si2 (oxime-cured, filled) was chosen to represent the silicones for testing in joint form, using as primer the recommended silane/acrylate. The detailed results of Appendix A5 were again (see Section 4.3) used in the form of modulus change ($\Delta\sigma_{25}$) and change in maximum extension (ΔE_{max}), referred to the wet control as standard. Also, for comparison, the same values were calculated from the corresponding strip test results, again using the wet control as standard. All of these values are presented in Table 4.6, below.

Table 4.6 *Changes in (a) 25 per cent modulus ($\Delta\sigma_{25}$) and (b) maximum extension (ΔE_{max}), relative to the wet controls, for (a) Si2/acrylate joints and (b) Si2 sealant strips exposed to various environmental conditions. (N = no specimens received or specimens not identified)*

(a) $\Delta\sigma_{25}$ *relative to wet control*

	Time point	Dry	Soil	Yorks pre-	Yorks post-	Tewks pre-	Tewks post-	Sew 99%	Sew out	Sew anaer
Si2 strips	1	+25.0	+112.5	+50.0	+37.5	+50.0	-37.5	+75.0	+75.0	+75.0
	2	-20.0	+15.0	+35.0	+30.0	-25.0	-60.0	+20.0	-20.0	+20.0
Si2/acrylate joints	1	+11.5	+1.5	0	-5.0	-10.0	+6.5	+31.5	+3.5	N
	2	+11.5	-2.8	-5.0	-9.6	-16.5	-3.7	-16.1	+0.9	-9.6

(b) ΔE_{max} *relative to wet control (per cent)*

	Time point	Dry	Soil	Yorks pre-	Yorks post-	Tewks pre-	Tewks post-	Sew 99%	Sew out	Sew anaer
Si2 strips	1	+36.8	+17.3	-16.1	-25.1	-2.7	-68.0	-45.7	-31.0	-14.5
	2	+142.0	+157.0	+147.5	+180.6	+49.4	-18.0	+183.2	+100.8	+122.5
Si2/acrylate joints	1	-19.2	-8.3	+10.9	-20.7	-25.9	-15.5	-10.9	-9.8	N
	2	+31.6	-45.9	+8.2	+12.2	+8.2	+30.6	+21.4	-37.8	+37.8

4.7.1 Comparative performance of joints after different environmental exposures

Examination of the detailed results underlying Table 4.6 indicates that the wet controls for the joints are quite reliable, replicates being confined to a narrow band and all showing cohesive failure. Generally speaking, the environmental test results are also similarly reliable and show cohesive failure. Thus, it is possible to compare results from the different environments with some confidence.

At the first time-point the modulus values are rather erratic generally, but the *Sew 99%* result in particular is abnormally high (20 per cent higher than the dry control, in fact). *Tewkes pre* is also lower than expected at -10 per cent. At the second time-point, all values except *Dry* and *Sew out* are negative, with *Sew 99%* and *Tewkes pre* showing the lowest values. No reason can be seen for the rather large fluctuations in the *Sew 99%* figures, since uniform cohesive failures were seen.

Except for *Yorks pre*, extensibility values at the first time-point are all negative compared with the wet control. At the second time-point, all values have become positive except for *Soil* (a sharp reduction here) and for *Sew out*. These results are generally consistent with a softening, hence increase in extensibility, between the two time-points, but no marked biodegradation. The general softening may be due to further uptake of water by the Phase 2 process.

4.7.2 Comparison of strips and joints after different environmental exposures

It is clear from a comparison of the joint and strip results that, although the absolute values of $\Delta\sigma_{25}$ and AE_{max} show large differences between the two cases, the same general trend underlies both sets of results. Thus, there is a reduction in modulus change and an increase in extensibility change between the two time-points, again suggesting a softening, possibly due to further water absorption. The dry controls, however, also give the same impression. Since this cannot be due to further curing, it must be caused by experimental variability, either in the wet or dry control tests.

As has been noted earlier, the results for *Tewkes post* stand out as being the worst in the strip tests. This can be seen in Table 4.6, where the extensibility changes are all negative and at much lower values than in the remaining test environments. However, this trend is not seen in the joint test results, where the performance of *Tewkes post* can only be described as average.

Thus, the strip tests predict a degradation that is not seen in the joint tests. However, it is expected that surface degradation effects will be magnified in the strip results due to the larger surface/volume ratio. Clearly, further testing would be required to determine whether joints would suffer the same degradation in the long term; if so, the strip results could be considered very valuable in the early prediction of biodegradation.

4.8 WIS RATINGS AND FAILURE MODES OBSERVED WITH SEALANT JOINTS

Summary Tables A5.5 and A5.6 show the WIS zone rating for environmentally exposed joint specimens using, in turn, the dry and wet control specimens as a reference datum point (refer to Appendix A1, Figure A1.5).

In the majority of cases, including the wet controls, the WIS classification against the dry controls is poor (l/o/u). However, when the wet controls are used as the datum point, the WIS classification is often the same as that given against the dry controls, or better. Consequently, the differences between wet and dry conditions are more marked than those between wet conditions and microbiologically active environments. This is particularly noticeable for the PSI/epoxy combination (and may also be the case for the PSI/isocyanate system).

These observations would appear to confirm the importance of establishing good performance under simple wet conditions before considering product systems for service in real environments. This requirement is further emphasised by the very high incidence of adhesive failure that was observed in this work, even for many of the wet controls at the first time-point. Exceptions are:

- Si2/acrylic, where a cohesive failure mode was found under virtually all exposure conditions
- PU3/epoxy, where cohesive failure was observed at the second time-point in four environments.

In some other cases, E_{max} remained reasonably close to the value given by the dry control despite the change from a cohesive to an adhesive failure mode. Thus, with these systems, acceptable service might be obtained in certain circumstances. Although, in this work, a distinction was not made between failure at the substrate/primer and the sealant/primer interfaces, the frequent occurrence of adhesive failure would suggest that further development and evaluation of primers for wet conditions is necessary.

4.9 MICROBIOLOGICAL INTERPRETATION

A major limitation of this study was the small number of time-points taken – one at three months and the other at six months. For adequate observation of trends in the selected performance indices more time-points are required. However, from the database obtained, the effects of microbiological challenges (both laboratory-based and environmental) upon the various sealant formulations can be ascertained.

4.9.1 Comparison of σ_{25} data for sealant strips challenged with different consortia and environments

The summary data presented in Tables 7.1, 7.2, 7.3 and 7.4 are as follows:

- Table 7.1. A positive change between the wet control and the six-month time-point reflects a softening of the sealant whereas a negative change reflects a stiffening. The changes between the three- and six-month time-points are readily apparent from data plotted, as in Figure 3.3 (Section 3.4.1).

    Table 7.1 presents a summary of $\Delta\sigma_{25}$ for each consortium challenge. These changes may, however, be a consequence of factors other than microbiological activity. For example, σ_{25} for the dry control should always be above that for the wet control; a reversal is an indication of continued curing of the sealant

- Table 7.2 is a summary of data obtained from the selected environmental locations and analysed as for Table 7.1

- Table 7.3 and Table 7.4: in addition to summarising the data for each consortium or environmental challenge, the data has been analysed with respect to each sealant formulation and is presented in these tables.

Changes in mechanical performance indices can justifiably be assumed to be of microbiological origin when comparison is made between the wet control, exposed to a physico/chemical but microbiologically inactive regime, and the exposed time-point sample.

It is difficult to make substantive microbiological conclusions from the above data in isolation.

These data tables are provided primarily for reference purposes. Future sealant formulations and challenge regimes may be compared with the performance of specific formulations in this research – reference recommendation at 6.4.

4.9.2 Comparison of the surface topology of sealant strips challenged with different consortia and environments

- Tables 7.5 and 7.6. The effect of the various challenge regimes on the surface topology of the sealant formulations is summarised in Table 7.5 for the consortia and in Table 7.6 for the environments.

It is believed that these changes are primarily a consequence of microbial activity and that the classification categories reflect the degree to which the sealant has been compromised by microbial activity. Again, these data have been referenced back to the wet controls exposed to identical challenge regimes while excluding the possibility of microbial activity.

The severity of sealant compromise follows two modes.

1. Holes progressing to a "sponge-like" consistency. From the available data this has been interpreted as loss of filler followed by microbial attack of the polymer *per se*.
2. Surface cracks progressing to surface crumbling. This effect is assumed to be a consequence of hydration changes in the sealant followed by degradation of polymer constituents in a progressive manner from the surface inwards.

It was anticipated that by comparing the data from Table 7.1 with that from Table 7.5, and from Table 7.2 with that from Table 7.6, correlation would be apparent as follows:

- the appearance of holes/sponge-like morphology would correlate with a softening of the sealant
- the appearance of cracks/crumbling would correlate with hardening of the sealant.

Such correlations are not readily apparent from the data.

4.9.3 Comparison of the effects of microbial consortia with those of selected environments on sealant strips

One of the main objectives of the work was to establish a correlation between selected microbial consortia and specified environmental conditions. This is not strictly possible, as the variances explicit in an environmental situation (chemical, physical and biological) cannot adequately be simulated in a simple laboratory test.

It must be recognised that it will not be possible to mimic categorically any environmental situation in the laboratory. The potential variables are too great and most will impact upon the sealant or the extent of microbial degradation.

However, comparison of σ_{25} values for sealant strips after six months with the wet controls for consortia and environmental challenges (see Appendix A3) do show a degree of correlation (Tables 7.7 and 7.8). Specific consortia, for example in Table 7.7, for sealant PU6 the effect of Consortia 10 (C10) is similar to that found for environmental exposure at *Tewkes post*. Comparisons of this nature can be made across and between these tables to gain an indication as to which consortia most adequately reflects the impact of a particular environment on a specified sealant formulation.

It must be emphasised that a degree of caution must be exercised in extrapolating from these data. There were insufficient data points analysed for trends in effects to be ascertained accurately. The fundamental basis of much of the data requires further in-depth analysis at the microbiological level.

Despite the above, the data presented lays a foundation for refining the test protocol presented and gives a large data reference base for future comparison both in terms of sealant formulations, challenge environments and microbial consortia (Appendix A3).

Table 4.7 presents a correlation between consortia and environmental challenges for each sealant type.

Table 4.7 *Correlation between consortia (C1 to C11) and environmental challenges for each sealant type. These data must be considered in conjunction with that presented in Tables 7.3, 7.4, 7.5 and 7.6. Only polymer formulations close to the commercial formulation have been considered in this table*

	Environment	Significant consortia (C)		
		PU	PS	Si
E1	Sewage outfall	4,5,6,7,8,9,10,11	1,2,3,5	4,5,6,7,8,9,10,11
E2	Sewage anaerobic 99% humidity	4,5,6,7,8,9,10,11	4,6,7,8,9,10,11	4,5,6,7,8,9,10,11
E3	Sewage anaerobic	4,5,6,7,8,9,10,11	1,2,3,5	4,5,6,7,8,9,10,11
E4	Raw water (upland)	4,5,6,7,8,9,10,11	4,6,7,8,9,10,11	4,5,6,7,8,9,10,11
E5	Upland treated water	4,5,6,7,8,9,10,11	1,2,3,5	4,5,6,7,8,9,10,11
E6	Raw water (lowland)	4,5,6,7,8,9,10,11	1,2,3,5	4,5,6,7,8,9,10,11
E7	Lowland treated water	4,5,6,7,8,9,10,11	1,2,3,5	4,5,6,7,8,9,10,11
E8	Soil	1,2,3	12,3,5	4,5,6,7,8,9,10,11

5 General discussion

5.1 INTRODUCTION

Section 4 presents individual results and discusses their interpretation discussed. This section addresses questions of importance to sealant manufacturers and users, as below.

1. What generic type of sealant polymer best resists microbiological attack under specific conditions?

2. What additional sealant components (eg, curatives, plasticisers) are best and which should be avoided?

3. What are the significant parameters that should be tested to evaluate or confirm the suitability of a particular formulated sealant for use in a specific field environment?

4. How should the tests be carried out?

5. To what extent is it possible to simulate a specified environment in the laboratory using microbial consortia?

6. How important is it to ensure that the sealant is fully cured before exposure?

Most of these queries are addressed in the sections that follow.

5.2 STATE OF CURE OF SEALANT BEFORE IMMERSION

In general it is considered desirable that a sealant should be fully cured before any testing of its performance is begun. It is fairly easy to asses the state of cure of a sealant in a test joint by means of a modulus test either in extension or compression. The latter was used successfully by Aubrey (1992) in tests involving water immersion.

In the present work, no specific testing for state of cure was carried out. Instead, the approach taken was to allow the sealant to cure for a longer time and at a higher temperature than those recommended by the manufacturers. It has been assumed that the sealant would be fully cured after this treatment.

The results of the present work, however, show that some of the polyurethanes are certainly not fully cured by the treatment given. The reasons for this difficulty with polyurethanes have been discussed (Section 4.2.1).

The polysulfides, in contrast, seem to be fully cured in the case of joints (Section 4.5) and almost fully cured in the case of strips (Section 4.4.1). The silicone strips show some evidence of further curing (Section 4.7.1), but it was not clear whether or not the joints were fully cured because of the erratic nature of the results (Appendix A5).

It must be significant that both the polyurethane and the silicone sealants are water-sensitive and the cure attained in the dry state must depend strongly on the atmospheric conditions prevailing. Once immersed, the rate of cure will be different and the ultimate extent of cure may well be different. The effects of this on the rate of development of cure (assessed by modulus) has been discussed in Section 2.1.5. This caused difficulties in the present study in distinguishing between modulus changes due to further cure and those due to water absorption or microbial degradation.

As the cure of a moisture-curing sealant commonly continues to give measurable modulus changes for several weeks or even months, it is not very practicable to prepare fully cured strips or joints for tests involving the detection of rather small modulus changes due to microbial activity. It should still be possible to see such changes, however, if the wet controls and microbially exposed samples are exposed side-by-side and at the same time in the same water with the control being sterile. It is therefore recommended that great care be taken to give the wet controls a treatment that is as identical as possible to the test samples, whether joints or strips are involved.

5.3 FIELD AND LABORATORY EXPOSURE TO MICRO-ORGANISMS

An objective of this study was to devise and evaluate microbial consortia that could be used in the laboratory to simulate the impact of a specified environment on a defined sealant formulation.

5.3.1 Selection of particular consortia and environments

By reference to published literature, micro-organisms that had been shown or inferred to have a role in polymer degradation were obtained from the National Collection of Industrial and Marine Bacteria (NCIB) or the American Type Culture Collection (ATCC). These were assembled into consortia depending on their physiology and environmental niche (Appendix A3).

In addition to the above, microbial isolates were obtained from sites where microbial degradation of sealants was apparent or suspected. The selected environmental sites were those most appropriate to the study, eg water and sewage treatment plants, or which would be expected to provide a strong microbial challenge.

5.3.2 Correlation between field and laboratory exposure

Recognising the limitations discussed in Section 5.3.1, it has, however, been possible to accumulate data that permits comparison of consortia, environments and sealant formulation. These data are presented in Tables 7.1, 7.2, 7.3, 7.4, 7.5 and 7.6. It is not possible to go beyond the interpretation of these data as presented in this document. However, this database does represent a solid reference source for future work.

Correlation between consortia and the selected environments is in part possible using Tables 7.7 and 7.8. However, as emphasised elsewhere, these data should be interpreted with a high degree of caution.

As discussed in Section 4.9 it must be recognised that it will not be possible to categorically mimic a particular environment by use of contrived microbial consortia. The environmental variations are too great and too many, and all will impact on the sealant to greater or lesser extent. There can be no adequate substitute for exposure of sealant to its intended environment.

5.4 EFFECTS OF SOME SEALANT VARIABLES

5.4.1 Comparison of sealant generic polymer types

It is of some interest to examine whether different generic types of polymers used in sealants are more susceptible than others to general microbiological degradation or to degradation by specific types of organism.

The simple formulations that approximate most closely to each polymer type used in sealants are PU8 (polymer only – polyether), PU9 (polymer only – polyester), PS5 (simple polysulfide) and Si1 (oxime-cured no filler). A comparison of these should therefore indicate the relative ease of attack by microbial agents on the polymer itself. More complex formulations are represented by PU1 (polyester) and PU3 (polyether).

Unfortunately, such a comparison is not very revealing. So far as the consortia results (Appendix A4) are concerned, the composition PU8 (polymer only – polyether) performs quite well under all conditions tested (except for the anomalously high first time-point values for C1–C3 and low values thereafter). Composition PU9 (polymer only – polyester) is somewhat inferior in its general performance, having some values of σ_{25} and E_{max} that are noticeably lower than those of the wet control. Composition PS5 (simple polysulfide) is also slightly inferior in its performance, with an anomalously high C10 modulus and erratic results generally. Composition Si1 (no filler) shows reasonably good performance after the second time-point has been reached, with consistently high $\sigma 25$ and E_{max} values, the only notable adverse changes being in reduced C9 and C11 E_{max} values.

In the environmentally exposed tests, PU9 (polymer only – polyester) is now more noticeably inferior to PU8 (polymer only – polyether). Most of the σ_{25} and E_{max} values are greatly reduced, the *Tewkes pre* and *post* values being particularly bad. PS5 (simple polysulfide) shows mainly increases in modulus compared with the wet control at both time-points, although E_{max} values are very variable. Si1 (no filler) shows generally good results throughout, with σ_{25} and E_{max} values at both time-points indicating a general stiffening compared with the wet control.

The general impression gained from these results is that both laboratory and field tests generally rank the polymers in the order: Si1 > PU8 (polyether) > PS5 > PU9 (polyester). The poorer performance of the ester-containing polymer PU9 is not so surprising if hydrolysis is considered an important mechanism. *It is emphasised that this impression is derived only from the limited data from this study, and is not intended to indicate in any way a universal trend. Had more polymers of each type been studied, the rankings might have been quite different.*

The fully formulated sealants PU1 (polyester) and PU3 (polyether) are identical in composition except for the type of polyol used. Examination of the results (Appendices A4 and A5) show no differences of consequence in the consortia tests and only marginal superiority in the case of PU3 (polyether) in the environmental tests.

5.4.2 Chlorinated versus phthalate plasticisers

It is presumed from general observations that chlorinated plasticisers should be more resistant to attack than phthalate plasticisers. In this work, direct comparisons of PU3 (chlorinated plasticiser) versus PU6 (phthalate plasticiser), as well as of PS2 (chlorinated plasticiser) versus PS3 (phthalate plasticiser) are provided in Tables 7.1–7.6.

Comparison of the results from consortia tests shows the following. There is very little difference between the results for the polyurethanes PU3 and PU6. By the second time-point, both show a general reduction in modulus compared with the wet control. E_{max} values are erratic, but generally lower in the C7–C11 range. The only case where PU6 is markedly inferior to PU3 is in the C10 case, where both σ25 and E_{max} values are very low. With PS2 and PS3 moduli show a moderate increase generally compared with the wet control, and E_{max} values are respectable after six months, although somewhat erratic. The consortia tests, therefore, do not reveal any striking differences between the sealants containing the two kinds of plasticiser.

Examination of the environmental test results for PU3 (chlorinated) and PU6 (phthalate) show, at the six-month time-point, a general increase in modulus and a reduction in extensibility compared with the three-month values. However, there are no obvious marked differences between the two types of sealant. In both cases the results for *Tewkes post* are notably poor, but this is obviously not connected with the plasticiser type. The results for PS2 (chlorinated) and PS3 (phthalate) were examined in Section 3.2.2, where they were seen to be very erratic but with fewer failures in the chlorinated (PS2) case. If we confine the examination to the second time-point, it is seen that, whereas most of the results for PS2 are still measurable, most of the PS3 results, for both modulus and extensibility, have declined to unmeasurable levels. Two obvious exceptions to this decline are the *Soil* and *Yorks pre* samples, which still appear to be in good condition in both cases. Nevertheless, it does appear that from this comparison there is clear evidence that the substitution of chlorinated oil by phthalate oil has resulted in inferior resistance to biodegradation generally.

5.4.3 Effects of heavy metal catalysts/curatives

The metals of relevance here are tin and mercury in the polyurethanes, and manganese and lead in the polysulfides. Although there is also tin in the silicones, there is no sample without it for comparison. The tin and mercury in the polyurethanes are in the form of organometallic catalysts that may remain unchanged, whereas the manganese and lead in the polysulfides are curatives, which will certainly change chemically during cure. Moreover, the nature of the cross-links also depends on the curative used in polysulfides (Hanhela *et al*, 1986), so that microbial attack may not simply depend on the metal-containing residues present.

Polyurethanes

Polyurethane PU3 (standard with tin catalyst) differs from PU5 (mercury catalyst) and PU4 (amine catalyst) only in the kind of catalyst used. Thus, the two metal-containing ones should be comparable directly with each other, and both may be compared with the non-metal-containing PU4.

The consortia tests reveal very little difference generally between these three samples. For example, the results for second time-point modulus σ_{25} show:

- PU3 – wet control 0.56 Mpa; mean of all consortia 0.45 Mpa
- PU4 – wet control 0.57 Mpa; mean of all consortia 0.49 Mpa
- PU5 – wet control 0.75 Mpa; mean of all consortia 0.59 Mpa.

The higher values for PU5 may merely reflect a greater rate of cure with the mercury catalyst. It may be significant, however, that the consortia results are far less evenly distributed in the PU5 case, where anomalously low values (0.16, 0.24, 0.23) are shown for three consortia (C8, C9, C10). Thus, if the mercury is protecting the sealant with respect to the other consortia it is certainly doing so with PU3, PU4 and PU5.

The environmental results are less revealing. The results for the sealants PU3 and PU4 are very similar in pattern and magnitude at both time-points. Sealant PU3 results are similar in overall pattern but again show generally higher σ_{25} values. In all three cases the *Tewkes post* environment produces strong signs of softening of the sealant, with no obvious protective effect by the mercury or tin.

Polysulfides

The polysulfides PS1 (manganese-cured) and PS2 (lead-cured) should provide a direct comparison of the manganese and lead curing systems. It should be borne in mind that the cross-link structures of these may differ as well as the type of metal (probably as oxide or salt) present.

The results for the consortia tests show that at the first time-point many σ_{25} results were very low, sometimes indistinguishable from zero. At the second time-point, however, the σ_{25} values are higher and follow a similar pattern for the two sealants. In every case the σ_{25} values are equal to or higher than those of the wet controls. There do not appear to be any catastrophically bad E_{max} values. The E_{max} values (second time-point) for PS2 (lead-cured) are higher generally than those of PS1 (manganese-cured), possibly indicating a stronger sealant. This may be more to do with the structures present than with microbial activity.

The results of the environmental tests are much more erratic but, if the second time-point results are considered, it can be seen that the σ_{25} values are generally lower than in the consortia tests, being unmeasurably low in some cases. This suggests that the environmental test is more severe than the laboratory one in degrading the sealants. It does not appear to be possible, however, to decide which sealant is performing better, in view of the erratic nature of the results. If the results for E_{max} are examined, however, it does appear that PS2 (lead-cured) has suffered more degradation than PS1 (manganese-cured), since the loss of extensibility compared to the wet control is generally greater.

5.5 SUITABILITY OF LABORATORY TESTING FOR PREDICTION OF SERVICE PERFORMANCE

As a result of this study, sealant manufacturers, specifiers and users will have at least five sources of information for reference when deciding on the suitability of a sealant for a particular field application:

1. Observation of the sealant in situ in service at the exposure site.

2. Removal and laboratory evaluation of test joints exposed to the site conditions.

3. Removal and laboratory evaluation of test strips exposed to the site conditions.

4. Evaluation of test strips exposed in the laboratory to consortia intended to simulate the exposure site (Section 4.9.3, Table 7.7 and 7.8 and Appendix A3).

5. Sealant manufacturers' literature and recommendations.

Except for (5), these various alternatives will be discussed in turn.

5.5.1 Scrutiny of sealant performance in situ

This is the definitive test of the suitability of a sealant or a joint for a particular application. However, it involves the difficulties and expense involved in scrutiny, or removal and replacement, as well as a long time-scale before meaningful results are obtained. The main purpose of the following tests is to avoid the need for this kind of evaluation.

5.5.2 Evaluation of test joints exposed at the test site

This would generally be considered the best means of testing (detailed in Section 3.1.2), in which the sealant/primer substrate model joint is exposed to the actual field environment to be used. It has the advantage that it allows any effects at the interfaces (sealant/primer or primer/substrate) to be detected as well as any microbial degradation of the sealant (or primer) itself. In the present work the test is, of course, a modification of the well-proven standard test (described in WIS 4-60-01 (1991)) for the performance of sealant joints in aqueous environments, extending it to their performance in microbially active environments.

The chief difficulty in using the test for the study of microbial degradation lies in distinguishing between the degradative effects of water only and water + microorganisms. Of course, this distinction can be ignored if a user is merely interested in whether or not a sealant/primer combination is suitable for exposure in a particular environment (when the unmodified WIS 4-60-01 is suitable).

The use of such joints in the present work for the unambiguous detection of microbial effects in the presence of water has not been very successful, as may be judged from the variability of results listed in Tables A5.1 and A5.2. It will also be apparent that the reliability of the wet control means is of the utmost importance in detecting trends in these tables. As has already been discussed in Section 5.2, the effects being observed in relation to the time-scale of three to six months are small, and the natural variability of the tests are too small to detect them. This reinforces the need for any future investigator to use more replicates, particularly of the wet control, and to test at more time-points extended over a longer period. To conform with such recommendations unfortunately involves much more expense, and the time and cost of making and testing so many extra joints may well be prohibitive. The avoidance of this time and expense is the prime reason for recommending the use of sealant strips instead of joints.

5.5.3 Evaluation of test strips exposed at the test site

The use of this test means sacrificing the ability to detect any interfacial effects, as it focuses simply on the effects of microbial attack on the sealant itself. The main advantages are (i) that the surface area/volume ratio is larger, hence effects should be detectable at an earlier stage, and (ii) it is much cheaper, with several hundred test joints typically being possible for the cost of one test joint.

Further benefits of the test specimens are that they can easily give information on the amount of water absorbed by sealants, and they provide controlled surfaces that can be examined microscopically to ascertain the nature of the microbial attack.

The test is most effective when used to follow modulus changes, which should be very reproducible for carefully cut test strips. In fact, variability of the order of ±10 per cent between replicates was routinely observed, probably reflecting variables in mixing, state of cure, porosity and surface flaws.

The many results obtained from this procedure can show statistically significant trends that should convince the user of the suitability of a particular sealant for use in service. Thus, in the polyurethane group, whereas the modulus values remain high generally, there is a distinct lowering of modulus (compared with the wet control) in the polyester products PU1 and PU9 (again suggesting hydrolysis) and a particularly drastic loss of modulus in the *Tewkes pre* and *post* exposures, indicating the unsuitability generally of polyurethanes for these environments (see Section 4.2.4 and Table 7.9). With the polysulfides (Section 4.4.4) there is a clear indication that the lead-cured samples are inferior to the manganese-cured samples insofar as retention of extensibility is concerned (the modulus values are too erratic to derive meaningful trends). With the silicones, there is again a clear degradative effect seen in the loss of modulus in the *Tewkes post* case. It is emphasised that these particularly obvious cases are identified from the relatively few results of the present work, whereas, work in the field for a single exposure site, involving more samples and time-points, would clearly be more definitive in ranking the different sealants tested. It is firmly believed that this type of testing, if done rigorously and comprehensively, would provide a very good basis for the selection of a sealant for a particular application.

5.5.4 Evaluation of test strips exposed to model microbiological consortia

This type of evaluation is amenable to much closer control of, for example, temperature, time, pH and microbial type, but its effectiveness for evaluating sealants depends mainly on whether or not the laboratory consortium chosen is a good simulant for the field site in question. This has been discussed in Section 4.9 and 5.3.

If a good simulant for the field site is assured, then this would naturally be the preferred and most useful type of test. Since the testing would be done entirely in the laboratory, the test would be free from the vagaries of the weather, seasons, or rainfall. It would be possible to control the challenge regime much more precisely and there would be the possibility of accelerating the degradative process by changing the temperature and/or the microbial concentrations.

However, the extent to which microbial consortia correlate with the actual test sites of the present study is not large. This has been discussed in Section 4.9 and 5.3 and can be appreciated by attempting to match the consortia and environmental results of Tables 7.7, 7.8 and 7.9. Although much more work is required and a definite procedure needs to be laid down to match test sites with laboratory consortia, the importance of the subject may well justify the initiation of this line of study by sealant manufacturers.

A deficiency of the present work was that an insufficient number of time-points were chosen for testing. In order to discern trends more readily, experience from this work has shown that several (say six) time-points should be taken at increasing intervals, for example, at one, two, four, six, nine and 12 months. It should be borne in mind that differences between test samples and wet controls will be small in the early stages and more replicates would be useful here.

During this present work it was observed that the range of modulus results was generally of the order ±10 per cent of the mean value. It maybe that the use of more replicates will enable this range to be more closely defined. The use of six time-points will enable a graph to be plotted, from which the gradient may show an initially insignificant change followed by a significant, unacceptable change, as shown in Figure 5.1.

At the same time-points, the surface of the samples should be examined by reflected light microscopy and, if possible, by scanning electron microscopy. The results of these examinations should be noted in the form described in Section 3.4.3. These will help to confirm the nature of the microbial degradation.

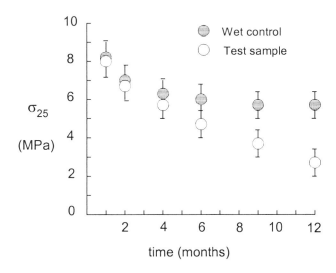

Figure 5.1 *Use of six time-points*

5.6 ESTABLISHING SUITABLE SEALANT FORMULATIONS FOR USE IN SPECIFIC ENVIRONMENTS

The question arises about the steps to be taken in sealant formulation or selection of sealants to obtain the best overall materials and procedures to ensure optimum resistance to microbial degradation. The above tests are designed to assess the suitability of only the sealant in resisting microbial degradation. The difficulties and expense involved in identifying microbial degradation with such tests have already been discussed in Section 5.5.2. It was also pointed out that the sealant user may simply require satisfactory performance in the standard WIS 4-60-01 (1991) test to ascertain the suitability of the joint to withstand immersion in the particular test liquid. Any microbial effects would then be combined with, but indistinguishable from, other aqueous degradative processes that would be shown in any analysis of the test results (Section 3.4.2).

If sealant products are being developed to resist identified microbiological conditions, or special projects require specific microbiological resistance, sealant strips should be subjected to challenge by selected microbial consortia as previously described (Section 4.9.3). This provides a screening process by which potentially suitable sealants can be identified and others eliminated. If necessary full environmental testing can follow.

The user may also need to check the suitability of a recommended primer, and that there are no adverse microbial effects at the interfaces. It should be pointed out that the present study provides little evidence for such degradation. Nevertheless, when selecting a primer the sealant user may require evidence based on, for example, the joint tests described in Sections 3.1.2 and 3.2.2 to confirm the absence of microbial attack on the primer and the associated interfaces.

6 Summary, recommendations and future work

6.1 ASSESSMENT OF THE SUSCEPTIBILITY OF SEALANT FORMULATIONS TO MICROBIAL DEGRADATION: TEST PROCEDURE

In the course of this study a test procedure for the assessment of the susceptibility of sealant formulations to microbial degradation has been devised (Appendix A1).

Previous studies, in particular the Water Industry Specification 4-60-01, utilised a test procedure based on mortar joints (Section 3.1.2). A major disadvantage of this test is the small sealant surface area accessible to microbial attack. As a consequence, if the assessment of compromise is by the performance indices used in this report, the susceptibility of the sealant to microbial activity would only become apparent after an extensive exposure time eg > 12 months. These difficulties are overcome in the sealant strip test (Appendix A1). However, in this no account is taken of the susceptibility of the sealant/primer/mortar interface to microbial attack (Section 4.8; Appendix A5). While this study provided little evidence of adverse microbial effects at joint interfaces, the high incidence of adhesion failures in the wet control samples indicates that strip tests should be complimented by wet joint tests to WIS 4-60-01 or ISO 13638.

Accepting the above, the sealant strip test has proved, nonetheless, to be a reliable and reproducible assay for the assessment of sealant performance when challenged environmentally or in the laboratory with defined microbial consortia.

It is important that in future studies to assess the susceptibility of specific sealants to microbial degradation more time-points are used, to enable trends and variations to be assessed. It is recommended that strip samples should be tested immediately before immersion and at monthly intervals for six months and then at one year.

6.2 ASSESSMENT OF THE SUSCEPTIBILITY OF SEALANT FORMULATIONS TO MICROBIAL DEGRADATION: MICROBIOLOGICAL CHALLENGE

The only definitive microbial challenge is immersion in the environment in which the sealant is to be used. However, this programme has derived a screening procedure by which potentially suitable sealants can be identified and others eliminated.

The reduction in challenge time, as a consequence of the increased exposed surface area of strips, makes this approach feasible and one which can be readily utilised by the sealant manufacturer.

Immersion of a sealant in an aqueous environment subjects the sealant to physical, chemical and biological challenge; the potential number of variables is considerable. To reduce these variables, so enhancing reproducibility of the assay, microbial mixtures (consortia) were prepared and used in laboratory-based challenge experiments to assess the susceptibility of selected sealant formulations to microbial degradation.

Challenging a sealant formulation with a microbial consortium broadly representative of a particular physiological group, and hence reflecting the microbiological activity of a particular environmental niche, was successful.

It is doubtful if a defined consortium will ever provide a degradative challenge that exactly represents the degradative potential of a particular environment (Tables 4.7, 7.7 and 7.8). An indication of susceptibility will be obtained, but it will not be possible to state categorically that the sealant will perform in the same fashion in a particular environment.

Adequate controls (sealant strip challenges under identical conditions but lacking microbial activity), enable data to be generated where the probability of any sealant change or compromise is a consequence of microbial activity.

6.3 ASSESSMENT OF THE SUSCEPTIBILITY OF SEALANT FORMULATIONS TO MICROBIAL DEGRADATION: EVALUATION

The performance indices chosen showed acceptable reproducibility.

The importance of microscopic observation of the sealant surface should not be overlooked. From the wide range of sealant formulations and environmental challenges only two modes of surface compromise were apparent: hardening, resulting in cracking/crumbling; and softening, leading to void formation and a spongy morphology.

It may well be that the surface topography will be a very early indicator of the susceptibility of a sealant to microbial degradation.

The fundamental principles behind these modes of microbial attack needs to be established if further progress is to be made with this approach.

6.4 ASSESSMENT OF THE SUSCEPTIBILITY OF SEALANT FORMULATIONS TO MICROBIAL DEGRADATION: SEALANT FORMULATIONS

In this study attention was given to a wide range of sealant formulations. Interpretation of these data is difficult, but this reference database will serve as a valuable foundation for future analyses of sealant formulations and their susceptibility to compromise by microbial degradation.

The importance of cure conditions on the performance of a sealant formulation requires further study. From the data presented in this report it can be inferred that inadequate cure can markedly affect the performance of the sealant. This conforms with more general observations relating other performance criteria to degree of cure.

6.5 ASSESSMENT OF THE SUSCEPTIBILITY OF SEALANT FORMULATIONS TO MICROBIAL DEGRADATION: JOINTS

As discussed above, the strip test does not present data pertinent to the stability of the sealant/primer/substrate interfaces to microbial challenge. From the data obtained it is apparent that the failure mode of the joint differs between challenges. It is not possible to judge whether this results from microbial degradation or a failure in joint preparation.

However, the joint test does not present a simple, rapid assessment assay for sealants. The strip test does.

6.6 FUTURE WORK

A procedure has been devised to investigate the ability of micro-organisms to degrade particular sealant formulations. However, the work presented does not address two important points: the effect of the degree of cure on sealant performance and primer degradation by micro-organisms.

6.6.1 Influence of the degree of cure on sealant performance

As has been indicated in several sections of this book, there is a question over whether or not some of the sealant formulations had undergone adequate cure. To evaluate fully the extent of microbial degradation on a sealant formulation it will be necessary to assess the impact of the degree of cure on performance.

6.6.2 Degradation of primers under wet conditions due to microbial attack

The test strip method does not provide a means of assessing the effect of microbial attack on primers. This is a very important aspect of joints that has not been addressed in this programme.

6.6.3 Development of a database of commercial sealants evaluated by the recommended test procedure

In addition to valuable data for both the manufacturing and user sectors of the industry, this would enable the procedure to be developed into a robust protocol.

7 Research data summaries

The following data were obtained as described in Section 3.4.1.

A positive change is indicative of softening whilst a negative change is indicative of hardening.

Table 7.1 Analysis of 20 per cent error bar charts of sealant strips exposed to defined microbial consortia for six months.

Table 7.2 Analysis of 20 per cent error bar charts of sealant strips exposed to different environments for six months.

Table 7.3 Polyurethane: analysis of 20 per cent error bar charts of sealant strips exposed to defined microbial consortia and environments for six months.

Table 7.4 Polysulfide and silicone: Analysis of 20 per cent error bar charts of sealant strips exposed to defined microbial consortia and environments for six months.

The following data were derived as described in Section 3.4.3.

Table 7.5 Summary of the surface morphology of sealant strips exposed to different microbial consortia as ascertained by light and scanning electron microscopy.

Table 7.6 Summary of the surface morphology of sealant strips exposed to different environments as ascertained by light and scanning electron microscopy.

Comparison of the σ_{25} values at six months compared to the wet controls for microbial consortia and environmental challenges.

Table 7.7 Comparison of the σ_{25} values for polyurethane strips after six months with the wet control for microbial consortia and environmental challenges. Plot of $[(\sigma_{25} \text{ wet} - \sigma_{25} \text{ sample T6})/\sigma_{25} \text{ wet}] \times 100$

Table 7.8 Comparison of the σ_{25} values for polysulfide and silicone strips after six months with the wet control for microbial consortia and environmental challenges. Plot of $[(\sigma_{25} \text{ wet} - \sigma_{25} \text{ sample T6})/\sigma_{25} \text{ wet}] \times 100$

Table 7.9 (a) σ_{25} data for sealant strips from both microbial consortia and environmental challenges after six months exposure.

(b) Comparison of the σ_{25} data for sealant strips after six months' exposure with the wet controls. Plot of $[(\sigma_{25} \text{ wet} - \sigma_{25} \text{ sample T6})/ \sigma_{25} \text{ wet}] \times 100$

Table 7.1 Analysis of 20 per cent error bar charts of σ_{25} data for sealant strips exposed to defined microbial consortia for six months

	CONSORTIA 1	CONSORTIA 2	CONSORTIA 3	CONSORTIA 4	CONSORTIA 5	CONSORTIA 6	CONSORTIA 7	CONSORTIA 8	CONSORTIA 9	CONSORTIA 10	CONSORTIA 11
POSITION (relative to wet controls at 6 months)											
ABOVE	PS1,PS2,PS3,PS4,PS5	PS1,PS2,PS4,PS5	PS1,PS2,PS3,PS4,PS5	PS1,PS2,PS3,PS4,PS5	PS1,PS2,PS4,PS5	PS1,PS3,PS4	PS1,PS2,PS3,PS4,PS5	PU9 PS1,PS2,PS3,PS4,PS5	PU9 PS1,PS2,PS3,PS4 PS5(1 time pt only)	PS1,PS2,PS3,PS4,PS5	PU8,PU9 PS1,PS3
	Si1,Si2,Si3	Si1,Si2,Si3	Si1	Si1,Si2	Si1	Si1	Si1	Si1,Si2,Si3	Si1,Si2,Si3	Si1,Si2,Si3	Si1,Si2,Si3
BELOW	PU1,PU2,PU3,PU4, PU5,PU6,PU7,PU8,	PU1,PU2,PU3,PU4, PU6,PU8,PU9	PU1,PU2,PU6,PU7, PU8,PU9	PU1,PU2,	PU1,PU2,PU6,PU8 PS3	PU1,PU2,PU6	PU1,PU2,PU3,PU6 PU9	PU1,PU2,PU4,PU5, PU6,	PU1,PU2,PU3,PU5, PU6,PU7,	PU1,PU2,PU5,PU6, PU9	PU1,PU2
INSIGNIFICANT CHANGE	PU9 Si1,Si2,Si3	PU5,PU7, PS3, Si1,Si2,Si3	PU3,PU4,PU5, Si2,Si3	PU3,PU4,PU5,PU6, PU7,PU8,PU9 Si2	PU3,PU4,PU5,PU7 PU9 Si2,Si3	PU3,PU4,PU5,PU7, PU8,PU9 PS2,PS5 Si2,Si3	PU4,PU5,PU7,PU8 Si2,Si3	PU3,PU7,PU8 Si1,Si2,Si3	PU4,PU8 Si1,Si2,Si3	PU3,PU4,PU7,PU8 Si1,Si2,Si3	PU3,PU4,PU5,PU6 PU7 PS2,PS4,PS5 Si1,Si2,Si3
CHANGE											
POSITIVE	PU6,PU7 Si3	PU1,PU2,PU3,PU4,PU5 PU5,PU6,PU7,PU8,PU9 PS1,PS2,PS3,PS4 PS5 Si1,Si2,Si3	PS4 PS1,PS2,PS3, PS5 Si1,Si2,Si3	PU1,PU2,PU3,PU5, PU6,PU7,PU8,PU9 PS1,PS2,PS3,PS4,PS5 Si1,Si2,Si3	PU1,PU2,PU3,PU4,PU5 PU6,PU7,PU8,PU9 PS3,PS4,PS5 Si1,Si2,Si3	PU1,PU2,PU3,PU4,PU5 PU6,PU7,PU8,PU9 PS1,PS2,PS3,PS4,PS5 Si1,Si2,Si3	PU1,PU2,PU3,PU4,PU5 PU6,PU7,PU8,PU9 PS1,PS2,PS3,PS4,PS5 Si1,Si2,Si3	PU1,PU2,PU3,PU4,PU5 PU6,PU7,PU8,PU9 PS1,PS2,PS3,PS4,PS5 Si1,Si2,Si3	PU1,PU2,PU3,PU4,PU5 PU6,PU7,PU8,PU9 PS1,PS2,PS3,PS4 Si1,Si2,Si3	PU1,PU2,PU3,PU4, PU5,PU7,PU8 PS1,PS3,PS4 Si1,Si2,Si3	PU1,PU2,PU3,PU4,PU5 PU6,PU7,PU8,PU9 PS1,PS2,PS3,PS4,PU5 Si1,Si2,Si3
NEGATIVE	PU1,PU2,PU3,PU4, PU5,PU8,PU9 PS1,PS2,PS3,PS4 PS5 Si1,Si2,Si3				PS1					PU6,PU9 PS5	
NOTES:					No PS2				PS5 1 POINT ONLY	No PS2	

Table 7.2 Analysis of 20 per cent error bar charts of sealant strips exposed to different environments for six months

		99& HUMIDITY	SEWAGE OUTFALL	SEWAGE ANAEROBIC	YORK. PRE-TREATMENT	YORK. POST-TREATMENT	TEWKES. PRE-TREATMENT	TEWKES. POST-TREATMENT	SOIL
POSITION (relative to wet controls at 6 months)	ABOVE	PU3,PU6,PU8		PU3,PU5,PU6,PU7,PU9	PU3,PU6	PU9			
		PS1,PS3(1 time pt only),PS4,	PS3,PS4,PS5	PS1,PS2(1 time pt only)	PS1,PS2,PS3,PS4,PS5	PS1,PS2,PS4	PS3(1 time pt only),PS4,PS5		PS1,PS2,PS3,PS4,PS5
				PS5					
		SI1,SI3	SI1	SI1,SI3	SI1,SI2,SI3	SI1,SI2,SI3	SI1,SI3	SI1	SI1,SI3
	BELOW	PU1,PU9	PU1,PU2,PU3,PU7	PU2	PU1,PU4,PU7	PU7	PU1,PU7,PU9	PU1, PU2,PU3,PU4,PI5, PU6,PU7,PU8,PU9	PU1,PU7,
			PS1,PS2,	PS4				PS1,PS5	
		SI2		SI2			SI2	SI2,SI3	
INSIGNIFICANT CHANGE		PU2,PU5,PU7	PU4,PU5,PU6,PU8,PU9	PU1,PU4,PU8	PU2,PU5,PU8,PU9	PU1,PU2,PU3,PU4,PU5,PU6,PU8	PU2,PU3,PU4,PU5,PU6,PU8		PU2,PU3,PU4,PU5,PU6,PU8,PU9
		PS2,PS5		PS4		PS3(1 time pt only),PS5	PS1,PS2	PS2,PS4	
			SI2,SI3	SI2					SI2
CHANGE	POSITIVE	PU1,PU2,PU3,PU5,PU6,	PU1,PU2,PU4,PU5,	PU1,PU3,PU5,	PU1,PU2,PU3,PU4,	PU1,PU2,PU3,PU4,PU5,	PU1,PU2,PU3,PU4,PU5,	PU1,PU2,PU3,PU4,PU5,	PU2,PU3,PU4,PU5,
		PU7,PU8	PU6,PU9	PU6,PU7,PU8,PU9	PU5,PU6,PU8,PU9	PU6,PU7,PU8,PU9	PU7,PU8, PU9	PU6,PU7,PU8	PU6,PU7,PU8
		PS1	PS3,PS4,PS5		PS1,PS2,PS3,PS4,PS5	PS2,PS4,PS5	PS5	PS2,PS4	PS2,PS3,PS4,PS5
		SI1,SI3	SI2,SI3	SI1,SI2,SI3	SI1,SI2,SI3	SI1,SI2,SI3	SI1,SI2,SI3	SI1,SI2	SI1,SI2,SI3
	NEGATIVE	PU9	PU3,PU7,PU8	PU2	PU7	PS1	PU6	PU9	PU1,PU9
		PS5	PS1,PS2	PS1,PS4			PS1,PS2, PS4	PS1,PS5	PS1
			SI1					SI3	
NOTES:		No PU4,		No PS3		ONLY 1 POINT FOR PS3	ONLY 1 POINT FOR PS3	No PS3	
		Only 1 pt for PU2, PS3, PS4 & SI2		ONLY 1 POINT FOR PS2 & PS5					

CIRIA C520

Table 7.3 Polyurethane: analysis of 20 per cent error bar charts of σ_{25} data for sealant strips exposed to defined microbial consortia and environments for six months

POSITION (relative to wet controls at 6 months)	PU1	PU2	PU3	PU4	PU5	PU6	PU7	PU8	PU9
above			H, SA, YPRE		SA	H, SA, YPRE	SA	C11	C8,C9,C11
								H	SA, YPOST
below	C1,C2,C3,C4,C5,C6 C7,C8,C9,C10,C11	C1,C2,C3,C4,C5,C6 C7,C8,C9,C10,C11	C1,C2,C7,C9	C1,C2,C8	C1,C8,C9,C10	C1,C2,C3,C5,C6,C7 C8,C9C10	C1,C3,C9	C1,C2,C3,C5	C2,C3,C7,C10
	H, SO, YPRE, TPRE, TPOST, S	SA, SO, TPOST	SO, YPOST	YPRE, TPOST	TPOST	TPOST	SO, YPRE, YPOST TPRE, TPOST, S	TPOST	H, TPRE, TPOST
insignificant change			C3,C4,C5,C6,C8, C10,C11	C3,C4,C5,C6,C7, C9,C10,C11	C2,C3,C4,C5,C6 C7,C11	C4,C11	C2,C4,C5,C6,C7 C8,C10,C11	C4,C6,C7,C8,C9 C10	C1,C4,C5,C6
	H, YPRE, YPOST TPRE, S	YPOST, TPRE, S	SO, SA, YPOST, TPRE, S	H, SO, YPRE, YPOST,TPRE,S	SO, YPRE, TPRE, S	H	SO, SA, YPRE, YPOST, TPRE, S	SO, YPRE, S	
change +	C4,C5,C6,C7,C8 C9,C10,C11	C4,C5,C6,C7,C8 C9,C10,C11	C4,C5,C6,C7,C8 C9,C10,C11	C5,C6,C7,C8 C9,C10,C11	C4,C5,C6,C7,C8 C9,C10,C11	C4,C5,C6,C7,C8 C9,C11	C4,C5,C6,C7,C8 C9,C10,C11	C4,C5,C6,C7,C8 C9,C10,C11	C4,C5,C6,C7,C8 C9,C11
	H, SO, SA, YPRE, YPOST, TPRE, TPOST	H, SO, YPRE, YPOST TPRE, TPOST, S	H, SA, YPRE, YPOST TPRE, TPOST, S	SO, SA, YPRE, YPOST TPRE, TPOST, S	H, SO, SA, YPRE, YPOST,TPRE, TPOST, S	H, SO, SA, YPRE, YPOST, TPOST, S	H, SA, YPOST, TPRE, TPOST, S	H, SA, YPRE, YPOST TPRE, TPOST, S	SO, SA, YPRE, YPOST, TPRE
−	C1,C2,C3	C1,C2,C3	C1,C2,C3	C1,C2,C3,C4	C1,C2,C3	C1,C2,C3,C10	C1,C2,C3	C1,C2,C3	C1,C2,C3,C10
	S	SA	SO			TPRE	SO, YPRE	SO	H, TPOST, S
NOTES:				NO H					

Table 7.4 Polysulfide and silicone: analysis of 20 per cent error bar charts of σ_{25} data for sealant strips exposed to defined microbial consortia for six months

		PS1	PS2	PU3	PS4	PS5	SI1	SI2	SI3
POSITION (relative to wet controls at 6 months)	above	C1,C2,C3,C4,C5,C6 C7,C8,C9,C10,C11	C1,C2,C3,C4,C5 C7,C8,C9,C10	C1,C3,C4,C6 C7,C8,C9,C10	C1,C2,C3,C4,C5,C6 C7,C8,C9,C10	C1,C2,C3,C4,C5 C7,C8,C9,C10	C3,C4,C5,C6,C7, C8,C9,C10,C11	C8,C9,C10,C11	C7,C8,C9,C10,C11
		H,SA, YPRE, YPOST, S	SA (1 time pt only), YPRE, YPOST, S	H (1 time pt only), SO, YPRE, TPRE(1 pt only), S	H, SO, YPRE, YPOST, TPRE, S	SO, SA, YPRE, TPRE, S	H, SO, SA, YPRE, YPOST, TPRE, TPOST, S	YPRE, YPOST	H, SA, YPRE, YPOST, TPRE, S
	below	SO, TPOST	SO	C5, YPOST(1 time pt only)		TPOST		H, TPRE, TPOST	TPOST
	insignificant change		C6,C11	C2,	C11	C6,C11	C1,C2	C1,C2,C3,C4,C5	C1,C2,C3,C4,C5
		YPOST (1 time pt only)	H (1pt only), TPRE, TPOST		SA, TPOST	H, YPOST		SO, SA, S	SO
CHANGE	+	C4,C6,C7,C8 C9,C10,C11	C4,C6,C7,C8 C9,C11	C4,C5,C6,C7,C8 C9,C10,C11	C3,C4,C5,C6,C7 C8,C9,C10,C11	C4,C5,C6,C7,C8 C9,C11	C4,C5,C6,C7,C8 C9,C10,C11	C4,C5,C6,C7,C8 C9,C10,C11	C4,C5,C6,C7,C8 C9,C10,C11
		H, YPRE,	YPRE, YPOST, TPOST,S	SO, YPRE, S	SO, YPRE, YPOST TPOST,S	SO, YPRE, YPOST, TPRE, S	H, SO, SA, YPRE, YPOST TPRE, TPOST, S	SO, SA, YPRE, S YPOST, TPRE, TPOST	H,SO, SA, YPRE, YPOST, TPRE, S
	-	C1,C2,C3,C5	C1,C2,C3	C1,C2,C3	C1,C2,	C1,C2,C3,C10	C1,C2,C3	C1,C2,C3	C1,C2,C3
		SO, SA, YPOST, TPRE, TPOST, S	SO, TPRE,		SA, TPRE	H, TPOST			TPOST
NOTES:			NO C5, C10			Only 1 point inC9			
			Only 1 point for H, SA	NO SA, TPOST only 1 pt for H,YPOST & TPRE	only 1 time pt for H, TPRE, S	Only 1 point inC9		only 1 time pt for H	

Table 7.5 Summary of the surface morphology of sealant strips exposed to different microbial consortia as ascertained by light and scanning electron microscopy

		Consortia 1	Consortia 2	Consortia 3	Consortia 4	Consortia 5	Consortia 6	Consortia 7	Consortia 8	Consortia 9	Consortia 10	Consortia 11
No change	1	PU2, PU3, PU7, PU8	PU2, PU3, PU4, PU7	PU1, PU2, PU4**, PU5	PU2, PU4, PU5, PU7	PU1, PU2, PU4, PU5**	PU2, PU4, PU6, PU7	PU2, PU3, PU4, PU5, PU6	PU2, PU3, PU4, PU5, PU7	PU1, PU2, PU3, PU4, PU5	PU2, PU3, PU4, PU5, PU7	PU1, PU2, PU3, PU4, PU5
			PU8	PU7, PU8	PU8, PU9**	PU7, PU8	PU8	PU8	PU8, PU9	PU6, PU7, PU8	PU8	PU6**, PU7, PU8
		PS2		PS2	PS1	PS1, PS2		PS2, PS3	PS3	PS2, PS3	PS1, PS2	PS2
		Si1	Si1	Si1, Si2, Si3*	Si1, Si2, Si3*	Si1, Si2, Si3	Si1, Si3*	Si1, Si2*, Si3	Si1, Si3*	Si1, Si2*, Si3*	Si1, Si2, Si3*	Si1, Si2*, Si3*
Holes	2	PU5*, PU6*, PU9**	PU1, PU5*, PU6*	PU3*, PU6*	PU1, PU3*, PU6*	PU3*, PU6*	PU1, PU3, PU5	PU1, PU7*	PU6		PU1, PU6*	PU9*
		PS1	PS1, PS2*	PS1, PS3*	PS2*, PS3*	PS3*	PS1, PS3*	PS1	PS1, PS2*	PS1**	PS3**	PS1*, PS3*
		Si2, Si3	Si1, Si3				Si2		Si2			
Spongy	3		PU1									
Cracks	4	PU1	PU9**	PU9**			PS2	PU9**				
							PU9			PU9	PU9	
Crumbly	5	PS3	PS3									
				Si1								
Fungal growth/ spores		PU1, PU2, PU7, PU9	PU1, PU2, PU7, PU9									
		PS2, PS3	PS2, PS3									
		Si2	Si1									
NOTES						No PU9				PS5 1 point only	No PS2	
						No PS2						

84   CIRIA C520

Table 7.6 Summary of the surface morphology of sealant strips exposed to different environments as ascertained by light and scanning electron microscopy

		99 per cent humidity	Sewage outfall	Sewage anaerobic	Yorks pre-treatment	Yorks post-treatment	Tewkes pre-treatment	Tewkes post-treatment	Soil
No change	1	PU3, PU5*, PU6**, PU7**, PU8 PU9	PU3*, PU8*	PU1*, PU2**, PU5**, PU8**	PU3, PU4, PU5**, PU6**, PU7	PU3, PU4, PU6, PU7, PU8	PU2, PU3*, PU4*, PU7*, PU8		PU2, PU6, PU7, PU8
					PS1, PS3	PS1, PS2	PS1		PS2
		Si1	Si1	Si1, Si2*, Si3*	Si1, Si2*	Si1	Si1	Si1	Si1, Si2*
Holes	2	PU1, PU2, PU4	PU1, PU2, PU4, PU6, PU7	PU3, PU4, PU7**	PU1, PU2	PU1, PU2, PU9**	PU1, PU6	PU1, PU2, PU3, PU5, PU6	PU1, PU3, PU4, PU5**
		PS1	PS1, PS2, PS3					PS1	
		Si2, Si3	Si2		Si3	Si2, Si3	Si2, Si3	Si2	Si3
Spongy	3	PU1, PU3	PU1					PU3, PU4, PU5, PU6, PU7, PU9	
		PS3		PS1	PS2			PS2	
			Si3					Si3	
Cracks	4		PU5, PU9	PU6, PU9	PU1, PU9	PU5	PU1, PU9		PU1, PU9
								Si2	
Crumbly	5	PU2					PU1, PU5	PU8	PU1
							PS2	PS1	PS1, PS3
		Si2				Si2			
Fungal growth/ spores		PU4, PU8		PU9	PU1, PU9	PU1, PU2, PU9	PU9		PU2, PU6, PU7
									PS3
		Si1	Si1			Si1	Si1, Si2		
NOTES		No PS2		No PS2, PS3		No PS3	No PS3	No PS3	

Table 7.7 Comparison of σ_{25} values for polyurethane strips after six months with the wet control for microbial consortia and environmental challenges. Plot of $[(\sigma_{25}\ wet - \sigma_{25}\ sample\ T6)/\sigma_{25}\ wet] \times 100$

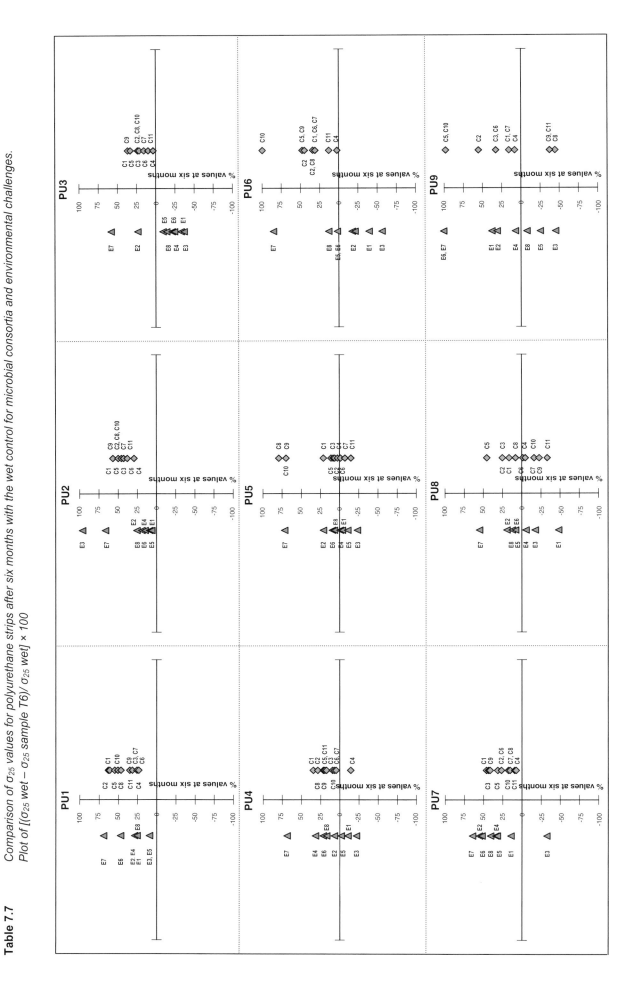

Table 7.8 Comparison of σ_{25} values for polysulfide and silicone strips after six months with the wet control for microbial consortia and environmental challenges. Plot of $[(\sigma_{25}\text{ wet} - \sigma_{25}\text{ sample T6})/\sigma_{25}\text{ wet}] \times 100$

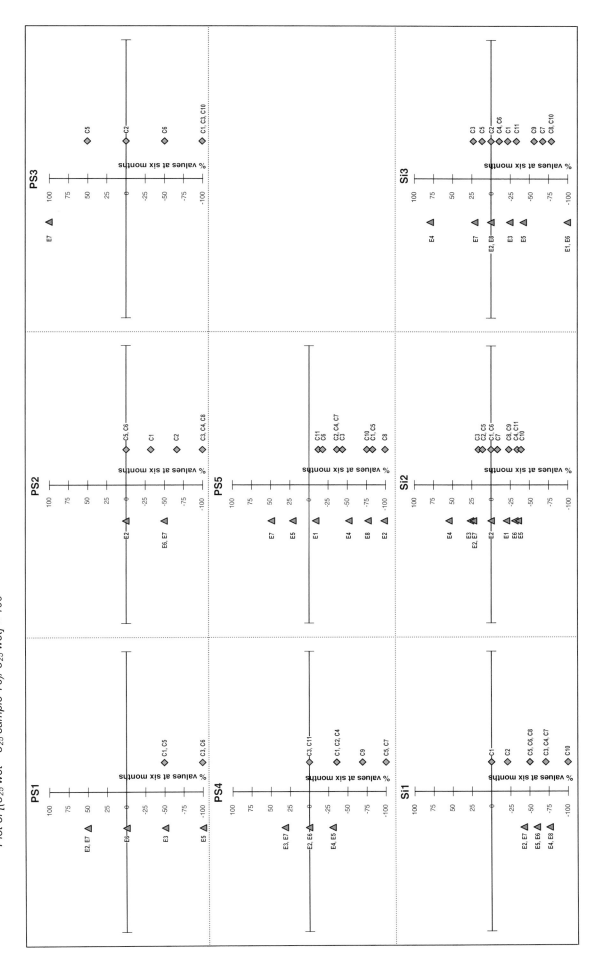

Table 7.9 (a) σ_{25} data for sealant strips from both microbial consortia and environmental challenges after six months' exposure
(b) Comparison of σ_{25} data for sealant strips after six months' exposure with wet controls. Plot of $[(\sigma_{25} \text{ wet} - \sigma_{25} \text{ sample T6})/ \sigma_{25} \text{ wet}] \times 100$

The highlighted values are those beyond the ±100 per cent scale presented in Tables 7.7 and 7.8

CONSORTIA – σ_{25} (MPa)

	PU1	PU2	PU3	PU4	PU5	PU6	PU7	PU8	PU9	PS1	PS2	PS3	PS4	PS5	SI1	SI2	SI3
dry	0.39	0.62	0.66	0.51	0.72	0.37	0.25	0.58	0.17	0.08	0.14	0.07	0.08	0.21	0.15	0.16	0.07
wet	0.57	0.87	0.56	0.57	0.75	0.4	0.33	0.6	0.13	0.02	0.02	0.04	0.03	0.16	0.1	0.2	0.09
consortia1	0.21	0.36	0.36	0.38	0.59	0.27	0.18	0.51	0.11	0.03	0.04	0.04	0.04	0.29	0.1	0.2	0.11
consortia2	0.2	0.43	0.43	0.44	0.7	0.22	0.24	0.45	0.06	0.05	0.05	0.02	0.04	0.22	0.12	0.18	0.09
consortia3	0.38	0.44	0.44	0.52	0.68	0.26	0.19	0.46	0.09	0.04	0.06	0.04	0.03	0.23	0.17	0.17	0.07
consortia4	0.42	0.6	0.55	0.65	0.73	0.39	0.31	0.62	0.12	0.05	0.06	0.05	0.04	0.22	0.17	0.27	0.1
consortia5	0.26	0.39	0.39	0.47	0.66	0.21	0.22	0.33		0.03	0.03	0.01	0.06	0.29	0.15	0.18	0.08
consortia6	0.44	0.5	0.5	0.55	0.77	0.27	0.24	0.61	0.09	0.04	0.03	0.03	0.08	0.19	0.15	0.2	0.1
consortia7	0.38	0.47	0.47	0.55	0.81	0.27	0.28	0.69	0.11	0.06	0.09	0.05	0.06	0.22	0.17	0.22	0.15
consortia8	0.31	0.43	0.43	0.41	0.16	0.26	0.28	0.56	0.19	0.06	0.06	0.08	0.07	0.32	0.15	0.25	0.16
consortia9	0.36	0.37	0.37	0.46	0.24	0.21	0.2	0.73	0.18	0.06	0.07	0.07	0.05		0.21	0.25	0.06
consortia10	0.28	0.43	0.43	0.54	0.23	0	0.3	0.7	0	0.07	0.07	0.04	0.08	0.28	0.2	0.28	0.16
consortia11	0.37	0.54	0.54	0.47	0.88	0.35	0.27	0.79	0.18	0.06	0.07	0.05	0.03	0.18	0.21	0.27	0.12

((WET CONTROL–SAMPLE)/WET CONTROL) ×100

	PU1	PU2	PU3	PU4	PU5	PU6	PU7	PU8	PU9	PS1	PS2	PS3	PS4	PS5	SI1	SI2	SI3
consortia1	63.16	58.62	35.71	33.33	21.33	32.50	45.45	15.00	15.38	-50.00	-33.33	-100.00	-33.33	-81.25	0.00	0.00	-22.22
consortia2	64.91	50.57	23.21	22.81	6.67	45.00	27.27	25.00	53.85	-150.00	-66.67	0.00	-33.33	-37.50	-20.00	10.00	0.00
consortia3	33.33	49.43	21.43	8.77	9.33	35.00	42.42	23.33	30.77	-100.00	-100.00	-100.00	0.00	-43.75	-70.00	15.00	22.22
consortia4	26.32	31.03	1.79	-14.04	2.67	2.50	6.06	-3.33	7.69	-150.00	-100.00	-150.00	-100.00	-37.50	-70.00	-35.00	-11.11
consortia5	54.39	55.17	30.36	17.54	12.00	47.50	33.33	45.00	100.00	-50.00	0.00	50.00	-166.67	-81.25	-50.00	10.00	11.11
consortia6	22.81	42.53	10.71	3.51	-2.67	32.50	27.27	-1.67	30.77	-100.00	0.00	-50.00	-100.00	-18.75	-50.00	0.00	-11.11
consortia7	33.33	45.98	16.07	3.51	-8.00	32.50	15.15	-15.00	15.38	-200.00	-200.00	-150.00	-100.00	-37.50	-70.00	-10.00	-66.67
consortia8	45.61	50.57	23.21	28.07	78.67	35.00	15.15	6.67	-46.15	-200.00	-100.00	-300.00	-133.33	-100.00	-50.00	-25.00	-77.78
consortia9	36.84	57.47	33.93	19.30	68.00	47.50	39.39	-21.67	-38.46	-200.00	-133.33	-250.00	-66.67		-110.00	-25.00	-55.56
consortia10	50.88	50.57	23.21	5.26	69.33	100.00	9.09	-16.67	0	-250.00	-133.33	-100.00	-166.67	-75.00	-100.00	-40.00	-77.78
consortia11	35.09	37.93	3.57	17.54	-17.33	12.50	18.18	-31.67	-38.46	-200.00	-133.33	-150.00	0.00	-12.50	-110.00	-35.00	-33.33

ENVIRONMENTAL – σ_{25} (MPa)

	PU1	PU2	PU3	PU4	PU5	PU6	PU7	PU8	PU9	PS1	PS2	PS3	PS4	PS5	SI1	SI2	SI3
dry control	0.39	0.62	0.66	0.54	0.58	0.37	0.24	0.58	0.17	0.08	0.14	0.07	0.08	0.21	0.15	0.23	0.12
wet	0.57	0.87	0.56	0.57	0.75	0.4	0.33	0.6	0.13	0.02	0.02	0.02	0.03	0.16	0.1	0.2	0.09
99% Humidity	0.44	0.79	0.72	0.63	0.79	0.56	0.29	0.89	0.08	0.07		0.18		0.18	0.27	0.24	0.18
Sewage outfall	0.42	0.65	0.44	0.53	0.58	0.47	0.15	0.49	0.09	0.01	0.02		0.03	0.32	0.14	0.16	0.09
Sewage anaerobic	0.51	0	0.74	0.69	0.93	0.64	0.43	0.7	0.19	0.03		0.18	0.02	0.42	0.21	0.15	0.11
York. pre-treatment	0.42	0.74	0.68	0.39	0.75	0.49	0.22	0.63	0.12	0.05	0.23	0.11	0.04	0.24	0.18	0.08	0.02
York. post-treatment	0.51	0.82	0.61	0.58	0.84	0.39	0.23	0.56	0.16	0.04	0.06		0.04	0.13	0.16	0.27	0.13
Tewk. pre-treatment	0.32	0.78	0.66	0.45	0.68	0.39	0.16	0.55	0	0.02	0.03	0	0.03	0.36	0.16	0.26	0.18
Tewk. post-treatment	0.18	0.31	0.25	0.18	0.22	0.07	0.12	0.27	0.14	0.01	0.03	0.33	0.02	0.08	0.14	0.16	0.07
Soil	0.43	0.7	0.63	0.48	0.7	0.35	0.2	0.53		0.08	0.26		0.16	0.28	0.18	0.2	0.09

((WET CONTROL–SAMPLE)/WET CONTROL) ×100

	PU1	PU2	PU3	PU4	PU5	PU6	PU7	PU8	PU9	PS1	PS2	PS3	PS4	PS5	SI1	SI2	SI3
99% Humidity	22.81	9.20	-28.57	-10.53	-5.33	-40.00	12.12	-48.33	38.46	-250.00	0.00	-800.00	0.00	-12.50	-170.00	-20.00	-100.00
Sewage outfall	26.32	25.29	21.43	7.02	22.67	-17.50	54.55	18.33	30.77	50.00			33.33	-100.00	-40.00	20.00	0.00
Sewage anaerobic	10.53	100.00	-32.14	-21.05	-24.00	-60.00	-30.30	-16.67	-46.15	-50.00	-1050.00	-450.00	-33.33	-162.50	-110.00	25.00	-22.22
York. pre-treatment	26.32	14.94	-21.43	31.58	0.00	-22.50	33.33	-5.00	7.69	-150.00	-200.00		-33.33	-50.00	-80.00	60.00	77.78
York. post-treatment	10.53	5.75	-8.93	-1.75	-12.00	2.50	30.30	6.67	-23.08	-100.00	-200.00		0.00	18.75	-60.00	-35.00	-44.44
Tewk. pre-treatment	43.86	10.34	-17.86	21.05	9.33	2.50	51.52	8.33	100.00	0.00	-50.00	100.00	33.33	-125.00	-60.00	-30.00	-100.00
Tewk. post-treatment	68.42	64.37	55.36	68.42	70.67	82.50	63.64	55.00	100.00	50.00	-1200.00	-1550.00	33.33	50.00	-40.00	20.00	22.22
Soil	24.56	19.54	-12.50	15.79	6.67	12.50	39.39	11.67	-7.69	-300.00			-433.33	-75.00	-80.00	0.00	0.00

8 References

ALBERTSSON, A C (1993)
JMS Pure Applied Chemistry A30 (9710): 757

ANON (1988)
"Joint Sealants in Concrete Structures in Wet Conditions"
Construction and Building Materials, Vol 2, No 3, pp 157–162

APPLETON, B (1973)
"Coming Apart at the Seals"
New Civil Engineer, 6 Dec 1973, p 8

AUBREY, D W (1992)
Performance of sealant-concrete joints in wet conditions: results of a laboratory testing programme. Volume 1: main results and discussion
CIRIA Technical Note 144

AUBREY, D W (1993)
Performance of sealant-concrete joints in wet conditions: results of a laboratory testing programme. Volume 2: schedules of detailed test results
CIRIA Project Report 8

AUBREY, D W and BEECH, J C (1989)
"The Influence of Moisture on Building Joint Sealants"
Building and Environment, Vol 24, no 2, pp 179–190

BEECH, J C and AUBREY, D W (1987)
Joint Primers and Sealants: Performance Between Porous Cladding
BRE Information Paper 9/87

BRIGGS, G J, EDWARDS, D C and STOREY, E B (1963)
"Water absorption of elastomers"
Rubber Chemistry and Technology, **36** p 621

BRITISH STANDARDS INSTITUTION
BS6093: 1993 *Code of Practice for Design of Joints and Jointing in Building Construction*

BRITISH STANDARDS INSTITUTION
BS6213: 1982 *British Standard Guide to Selection of Constructional Sealants*

CIRIA (1987)
Civil engineering sealants in wet conditions – review of performance and interim guidance on use
Technical Note 128

CIRIA (1991)
Manual of Good Practice in Sealant Application
Special Publication 80

DARBY, R T and KAPLAN, A M (1968)
"Fungal susceptibility of polyurethanes"
Applied Microbiology, **16,** p 900

DWYER, D F and TIEDJE, J M (1983)
Applied Environmental Microbiology, **46**, p185

EDWARDS, D C (1985)
"Water absorption phenomena in elastomers"
Elastomerics, Vol 117, no 10, October 1985, pp 25–30

FIELDS, R D, RODRIGUEZ, F and FINN, R K (1974)
J Applied Polymer Science, **18,** p 3571

GRANT, M A and PAYNE W J (1983)
Biotech Bioeng, **25**, p 627

HANHELA, P J, HUANG, R H E and BRENTON PAUL, D (1986)
"Water immersion of polysulfide sealants I"
Ind Eng Chem Prod Res Dev, **25**, pp 328–332

HANHELA, P J, HUANG, R H E, BRENTON PAUL, D and SYMES, T E F (1986)
"Water immersion of polysulfide sealants II"
J Appl Polym Sci, **32**, pp 5415–5430

KAPLAN, A M, DARBY, R T, GREENBERGER, M and ROGERS, M R (1968)
"Microbial Deterioration of Polyurethane Systems"
Developments in Industrial Microbiology, **9**, pp 201–207

KAWAI, F (1987)
CRC Critical Reviews Biotechnology, **6**, p 273

LEDBETTER, S R, HURLEY, S and SHEEHAN, A (1998)
Sealant joints in the external envelope of buildings: a guide to design, specification and construction
CIRIA Report 178

MANSFIELD, C (1990)
"Tests for the Water Resistance of Construction Sealants"
Construction and Building Materials, Vol 4, no 1, pp 37–42

SCOTT, G (1975)
Polymer Age, **6**, p 54

SEAL, K J and PATHIRANA, R A (1982)
Internat Biodeter Bull, **18**, p 81

SHINK, B and STIEB, M (1983)
Applied Environmental Microbiology, **45**, p 1905

TOKIWA, Y and SUZUKI, T (1977)
Nature, **270**, p 76

TOKIWA, Y and SUZUKI, T (1988)
Agricultural Biological Chemistry, **52**, p 1937

WATER INDUSTRY SPECIFICATION 4-60-01
Specification for building and construction joint sealants
WSA/FWR Sewers and Water Mains Committee: Materials and Standards (March 1991) Issue 1

WOLF, A T (1996)
"Ageing Resistance of Building and Construction Sealants (Part 1)"
In: Beech, J C and Wolf, A T (eds), *Durability of Building Sealants*, E & FN Spon, pp 63–89

WOOLMAN, R AND HUTCHINSON, A (1994)
Resealing of Buildings - A Guide to Good Practice
Butterworth-Heinemann Ltd

A1 Recommended test procedures

A1.1 SEALANT STRIPS

A1.1.1 Preparation of sealant strips

1. Sealant should be cast to a thickness of 1 mm over a prescribed area on a non-migatory silicone treated glass surface.

2. When sufficiently set a thin sheet of polyethylene should be applied to the exposed surface to prevent distortion.

3. Subsequently the composite sheet should be removed from the glass and a second sheet of polyethylene applied to the cast surface of the sealant.

4. The entire laminate should be cut into strips 10 × 100 mm.

5. Strips should be conditioned for at least 28 days at 35 °C to ensure adequate cure.

6. The protective polyethylene strips should be removed immediately before testing.

7. Prior to exposure to the prescribed test conditions the sealant strips should be weighed and their thickness accurately measured.

A1.1.2 Immersion procedures

Test strips should be prepared and immersed in the following ways.

1. Test strips should be immersed in 3 ppm chlorine solution for 10 minutes, rinsed thoroughly with sterile distilled water and placed in sterile, flat tissue culture bottles (Figure 3.1, Section 3.3.2). 50 ml of sterile distilled water (pH 6.8), 0.5 ml of a minimal salts solution and 5 ml inoculum should be added.

2. Test strips of each sealant formulation (three for each time-point) should be threaded onto nylon and suspended in a wire basket, which should then be sealed with plastic mesh (Figure 3.2, Section 3.3.3). The basket should be immersed in a 3 ppm chlorine solution for 10 minutes, rinsed thoroughly with water from the immersion site (or culture medium if a laboratory challenge is being undertaken) and then immersed at the chosen location or challenged with the selected consortia.

A1.1.3 Selection of microbial consortia

Table 4.7, Section 3.3, Section 4.9.3 and Appendix A3.

A1.1.4 Test method

The test procedure is based upon tensile testing using modulus and extensibility as the performance indices.

Strip samples should be tested immediately before immersion and thereafter at monthly intervals up to six months and at one year.

All sealant strips should be tested wet, except for the dry control, on a Davenport-Nene (or similar) tensile-strength-testing instrument at room temperature.

A standard 10 mm length, top and bottom of the test strip, should be secured in smooth callipers and positioned on the instrument.

Testing of each sealant type per exposure condition should be carried out in triplicate and the data averaged (Figure A1.1)

The recommended instrument settings are:

Load
Load cell rating   0.500 (kN)
Maximum load   0.200 (kN)
Load units   Mpa

Displacement
Method of measuring displacement (E/C/H)   C
Extensometer range   mm
Maximum displacement   800 mm
Displacement units   per cent
Gauge length   80 mm

Test mode
Test speed   50 mm/minute
Tension/compression/flexure   tension
Recorded data as (LD/LT/DT)   LD

In addition to tensile strength testing, the initial and final weights, thicknesses and lengths should be determined and recorded.

Stereoscanning electron microscopy and light microscopy should be carried out on test strips showing significant changes.

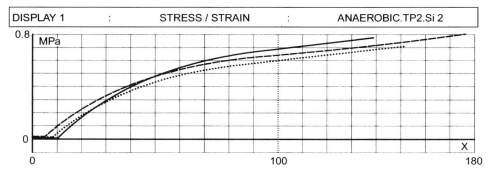

Figure A1.1 *Data from Davenport-Nene tensile strength testing instrument. Sealant strips were tested in triplicate and the results averaged*

A1.1.5 Evaluation of data

The results from a triplicate strip test should be averaged and an estimated standard deviation calculated. It has been found that the standard deviation falls within quite a narrow range (± 5–6 per cent of the mean figure) over many of the test strip triplicate results, suggesting that the overall errors inherent in the test itself could be reasonably predicted from this figure. Since, for a normal distribution, there is a probability of 0.95 that a value lies within two standard deviations of the mean, an error of no more than ±10 per cent of the mean value should be expected. This range should be calculated for every strip test triplicate result at the six-month time period.

Clearly, a comprehensive test for microbiological degradation of a strip would show a gradual change with time, and results should be taken at several time-points for this purpose. In the present work, this was not possible as it would have magnified the number of tests beyond the capability of the programme. Only two time-points were taken (three and six months), to see if consistent changes could be detected within this period. Thus, the first time-point value ranges (within ±10 per cent of the means) were compared with the second time-point value ranges (within ±10 per cent of the means) (Figure A1.2).

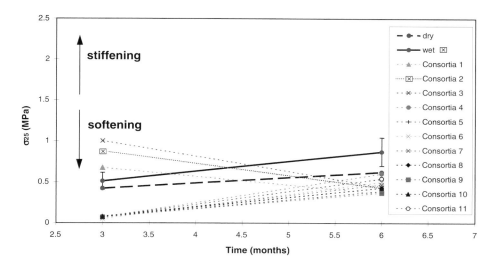

Figure A1.2 *Relationship between σ_{25} and the exposure time of PU2 to different microbial consortia*

If the first time-point values are less than the second time-point values (ie with no overlapping of the ranges of these results), this is termed a *positive change*. Conversely, if the first time-point values are greater than the second (with no range overlap), this is termed a *negative change*. If there is overlapping of the two ranges, there should be deemed to be no significant difference between the values obtained.

Of more significance in detecting a change as a consequence of microbiological attack, the first and second time-point values should be compared with the corresponding values for the wet controls. A positive change for modulus here indicates a hardening of the sealant from microbial attack, whereas a negative change indicates a softening. Typical changes are illustrated diagrammatically in Figure A1.3.

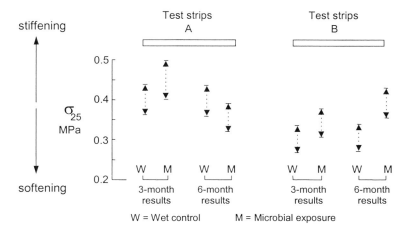

Figure A1.3 *Method of evaluating strip test results for σ_{25}*

In Figure A1.3, test strips A would be considered to show insignificant change from the wet controls at both time-points. However, whereas the wet control had shown no significant change between three and six months, the exposed sample had shown a significant decline in modulus (ie a negative change).

Test strips B were not significantly different from the wet controls at the first time-point, but showed a significantly increased modulus (ie a positive change) at the second time-point. Neither the wet controls nor the microbially exposed sample showed significant change between the three- and six-month time-points.

Clearly, further time-points would have made any trend in results more apparent.

(For reference these assessments of σ_{25} changes have been summarised in tabular form for all strip tests in Section 4.)

Scanning electron microscopy was found to be the most informative analytical procedure for assessing the effect of microbial activity on a sealant surface. However, this had to be supplemented with high power light microscopy to ensure that a sufficiently large surface area was evaluated for each test specimen. Such analyses were pertinent only to the sealant strips. Joints could not be adequately evaluated by these procedures.

Changes in surface morphology are categorised as follows (Figure 3.5; Table 7.5):

- No change: no perceived change when compared to the wet control
- Holes: holes apparent on the sealant surface
- Spongy: surface has an overall open, sponge-like, appearance
- Cracks: cracks, of varying depths and extent apparent on the sealant surface
- Crumbly: manifestation of extensive cracking of the surface.

The above, as listed, are perceived as being progressive effects of microbial degradation. They are, however, subjective and although of considerable value, reproducibility between different manufacturers and laboratories will be questionable without computerised image analysis.

In addition to microscopic analysis dimensional measurements should be taken (length and width) together with absolute weight.

A1.2 JOINTS

A1.2.1 Preparation and joint test protocol

The test joint and test procedure are as described previously (CIRIA: TN128 and PR8) and as published in the Water Industry Specification WIS 4-60-01 (1991).

The test specimen configuration is now very well known and is illustrated by the drawing in Figure A 1.4.

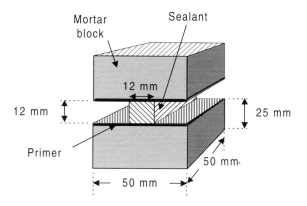

Figure A1.4 *Design of test specimen used for all joint tests*

To prepare the test joints moulds were assembled using spacers and release agents as recommended in the Water Industry Specification WIS 4-60-01 (1991).

In the present work, polypropylene-covered hardwood spacers were used. No additional release agents, which might transfer to the sealant surface, were used. Primer was applied to the mortar blocks and allowed to dry/cure for the prescribed time (following the manufacturers' instructions) before assembling the mould. Freshly mixed sealant was gunned into the assembled mould, then hand pressure was applied by means of a spatula to the ends of the sealant bead to displace air and ensure mould filling. The ends of the sealant bead were finally trowelled smooth.

The filled moulds were dismantled after 24 hours at 20 °C (three days in the case of the silicone sealants). The joints were then marked indelibly with identification numbers. To maintain a high rate of curing, the joints were placed in a fan oven at 35 °C, taking care to place the weight on the concrete substrates and not on the sealant beads. The joints were left to cure for 28 days, then cooled and stored dry for a day or two before sending to the appropriate immersion sites.

Two exposure times of approximately three and six months' duration were allowed for each test specimen type.

A1.2.2 Curing protocol

Joints should be stacked in a fan-assisted oven at 35 °C, taking care to avoid distortion of the sealant beads.

Joints should be left to cure for 28 days then cooled and stored dry (for a few days) before immersion at the test sites.

A1.2.3 Test method

This was as specified in WIS 4-60-01.

In the present work each set of joints (generally a minimum of three for each nominally identical sealant and exposure condition) was tested to failure in direct tension at 5 mm min^{-1} and 23 ±2 °C. This was carried out by Taywood Engineering Ltd using an Instron 1195 testing machine.

The recorded force in newtons at 25 per cent extension (F_{25}), the maximum force (F_{max}) and the extension at maximum force (E_{max}) were noted. The mode of failure was also noted after the test.

As in the WIS procedure, the processing of test results is simplified by utilising force rather than stress. This approach requires that the cross-sectional area of the sealant is essentially the same for all test specimens.

Further simplification is achieved by using the force at 25 per cent extension (F_{25}) in place of the secant modulus at this point. However, for a given cross-sectional area, there is a direct proportionality between these properties as this secant modulus is given by stress/strain at 25 per cent extension, ie stress at 25 per cent extension × 4.

The WIS procedure gives no guidance on the method that should be used to average replicate results when there are marked differences within the set. This is particularly likely to occur with variations in the failure mode. An adhesive failure, for example, can lead to a very much lower F_{max} and E_{max} in comparison to a cohesive failure, although F_{25} in each case may be very similar. In essence, changes at the sealant/substrate interface may show a lower repeatability than those that take place in the bulk of the sealant itself.

In the present work, F_{25} is, in general, the mean of all the results in a set; F_{max} and E_{max} refer to the predominant failure mode only. Thus, the bracketed results in the two summary tables (Appendix A5) are merely informative and have been discounted in the further analysis of the results. This procedure has been adopted throughout although, in a number of cases, F_{25} and E_{max} are reasonably close for different failure modes. Additionally, this procedure occasionally leads to F_{25} being greater than F_{max} (for example, PU6/epoxy from Stansfield View Storage Reservoir).

Together with the mean values for F_{25}, F_{max} and E_{max}, the summary tables referred to above also give the failure modes for the joint specimens. Additional summary tables, also in Appendix A5, give the percentage changes in modulus (ΔF_{25}) and extensibility (ΔE_{max}) relative to both the dry and wet control samples.

A1.2.4 Evaluation of data

The joint test results for ΔF_{25} and ΔE_{max} should be further interpreted using the method described in the Standard WIS 4-60-01, outlined below.

The percentage changes ΔF_{25} and ΔE_{max} should be used to assess a joint for acceptable performance by reference to the diagram shown below.

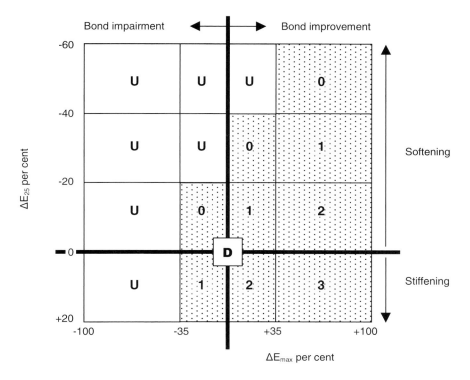

Figure A1.5 *Changes to modulus and extensibility after immersion (after WIS 4-60-01)*

In Figure A1.5, unshaded zones represent an *unacceptable* (U) change. Shaded zones represent a *potentially acceptable* change, the level of acceptability being indicated by the numbers 3 (most acceptable) to 0 (least acceptable).

In Water Industry Specification 4-60-01 the datum point refers to the dry control results, since this specification is concerned only with changes after water immersion. In the present work, this is still of relevance and WIS zone ratings for the joints tested, relative to the dry controls, are summarised in Appendix A5 (Table A5.5).

More importantly, in the present work, the results after microbial exposure are required to be compared with the wet controls. WIS zone ratings are therefore also presented in Appendix A5 (Table A5.6), using the wet controls as reference datum points.

By comparing the magnitude of changes induced by specific exposure regimes against both the wet and dry controls, it is possible to demonstrate the relative importance of microbial activity and water on the performance of the joint assembly.

In the summary tables given in Appendix A5, a WIS zone rating is given in some cases although F_{25} is not available due to premature failure. This is possible where ΔE_{max} is unacceptable, irrespective of the value of ΔF_{25}.

An implicit principle of the rating system outlined above is that bond integrity is a primary performance requirement. Hence, a significant increase in extensibility (ΔE_{max}) is given a positive rating. Positive changes in the extensibility are then ranked and unacceptable changes (U) identified by considering any softening of the sealant following immersion.

When an acceptable WIS zone rating is obtained, it is desirable that failure occurs in a cohesive mode. The occurrence of adhesive failures under test conditions usually suggests that premature failure is likely to be observed under equivalent (or more severe) service exposure.

The WIS describes two failure modes:

CF   cohesive failure in the sealant

AF   adhesive failure at or near the test surface.

More strictly, AF should be subdivided as follows:

(a)   adhesive failure at the test surface

(b)   thin film cohesive failure near the test surface

(c)   a mixture of adhesive/cohesive failure modes

For a given failure stress/strain, (b) is more acceptable than (a) or (c).

In the present work, a distinction was made between adhesive failure at the test surface ((a) above and coded A in the summary tables) and failure near the surface ((b) and (c) above and coded C/A in the tables).

As (b) and (c) were not separated, C/A may thus represent either a more or less acceptable mode of failure.

The use of C refers to cohesive failure in the bulk of the sealant.

A more precise system for classifying the failure mode would be:

A   adhesive failure at the interface

C   cohesive failure in the bulk of the sealant

TFC   cohesive failure in the sealant very close to the interface (thin-film cohesive failure).

Obliqued combinations of these abbreviations could then be used to indicate mixed failure modes. In some work, it might also be desirable to distinguish between substrate/primer and primer/sealant adhesive failure. With the risk of over-simplification, the WIS description does have the benefit of being easier to use in comparison with these alternatives.

A2   Sealant formulations

A2.1   SEALANT STRIPS

A2.1.1   Polyurethane

Table A2.1 *Sealant formulations: polyurethane sealant strips*

Designation	PU1	PU2	PU3	PU4	PU5	PU6	PU7	PU8	PU9
Desmophen 1652	21.0	–	–	–	–	–	–	–	21.0
Desmophen 1920D	–	21.5	22.2	22.2	22.2	22.2	22.2	22.2	–
Cerechlor 63L	26.0	26.0	26.0	26.0	26.0	–	26.0	–	–
Santicizer 160	–	–	–	–	–	26.0	–	–	–
Polestar 200R	13.0	13.0	13.0	13.0	13.0	13.0	–	–	–
Mellite 12	0.05	0.05	0.05	–	–	0.05	0.05	–	–
Dabco 33LV	–	–	–	0.10	–	–	–	–	–
Thorcat 535	–	–	–	–	0.05	–	–	–	–
Desmodur VLR20	3.0	–	1.6	1.6	1.6	1.6	1.6	1.6	3.0
Desmodur N3200	–	2.1	–	–	–	–	–	–	–
Isocyanate index	1.05	1.09	1.08	1.08	1.08	1.08	1.08	1.08	1.05
Hardness	49	54	51	48	54	–	38	62	–
IRHD (±2) (after seven days at 40 °C)									

A2.1.2   Silicone

Table A2.2 *Sealant formulations: oxime-cured silicone sealant strips*

Designation		Si1	Si2
Polymer	OH-terminated polydimethysiloxane	+	+
Plasticiser	Methyl-terminated polydimethysiloxane	+	+
Filler 1	Fumed silica	+	+
Filler 2	Calcium carbonate, stearic acid coated	–	+
Crosslinker	Methyl tris (butanonoximo) silane	+	+
	Tetrakis (butanonoximo) silane	+	+
Adhesion promoter	Amino-functional siloxane	+	+
Catalyst	Organotin (IV) carboxylate	+	+
Stabiliser	Ethylene oxide – propylene oxide copolymer	–	+

Key: + present   – not present

Table A2.3 *Sealant formulations: benzamide-cured silicone sealant strips*

Designation		Si3
Polymer	Dihydroxy polydimethyl siloxane	+
Plasticiser	Dimethyl polydimethyl siloxane	+
Catalyst	Dibutyl titanium diacetoacetic ester chelate	+
Filler 1	Calcium carbonate (calcite)*	+
Filler 2	Fumed silica	+
Catalyst	Dibutyl tin dilaurate	+
Curing agent	Arylamidoalkyl alkoxysilane	+

Key: * also contains low levels of titanium dioxide and carbon black   + present

A2.1.3 Polysulfide

Table A2.4 *Sealant formulations: polysulfide sealant strips*

Designation	PS1	PS2	PS3	PS4	PS5
Part A					
LP-32C	100	100	100	100	–
Thiokol ST	–	–	–	–	100
Cerechlor 63L	35	35	–	35	–
Santicizer 261	–	–	35	–	–
TiO_2 RCR2	10	10	10	10	–
Winnofil SPT	25	25	25	25	–
Polcarb S	30	30	30	30	–
Stearic acid	–	0.35	0.35	–	–
Sulfur	0.1	0.1	0.1	0.1	–
Thixatrol ST	–	–	–	5	–
Necires EPXL5	5	5	5	5	–
Zinc peroxide	–	–	–	–	5
Calcium hydroxide	–	–	–	–	1
Part B					
MnO_2	10	–	–	10	
PbO_2 MF	–	7.5	7.5	–	
Cerechlor 63L	12	4.5	–	12	
Santicizer 261	–	–	4.5	–	
Stearic acid	–	0.3	0.3	–	
TMTD	0.5	–	–	0.5	

Mix ratio A : B by weight		
PS1	205.1 : 22.5	
PS2 and PS3	205.4 : 12.3	
PS4	210.1 : 22.5	

A2.2 JOINT FORMULATIONS

Table A2.5 *Joint formulations*

	Joint type	
1	Polyurethane PU3 with two-part epoxy primer P13x	(P13 FOSROC EXPANDITE)
2	Polyurethane PU6 with two-part epoxy primer P13x	
3	Polysulfide PS1 with two-part epoxy primer W38	(W38 Morton)
4	Polysulfide PS3 with two-part epoxy primer W38	
5	Silicone Si2 with acrylate/silicone primer Arbo 2650	(Arbo 2650 Adshead-Ratcliffe)
6	PU3 sealant with isocyanate primer P20	(P20 FOSROC EXPANDITE)
7	PS1 sealant with isocyanate primer P20	

Note: Where the tin catalyst (Mellite 12) was included in joint formulations the content was reduced from 0.05 parts to 0.02 parts by weight

A3 Microbial consortia and challenge environments

A3.1 MICROBIAL CONSORTIA

Table A3.1 Microbial consortia

C1	*091856	*Aspergillus flavus*	C7		*Pseudomonas aeruginosa*
	319454	*A. fumigatus*			*P. alcaligenes*
	091855	*A. niger*			*P. aurantiaca*
	045545	*A. versicolor*			*P. aureofaciens*
	045550	*Chaetomium globosum*			*P. boreopolis*
	319459	*C. globosum*			*P. citronellolis*
					P. cocovenenans
C2	314385	*Fusarium solani*			*P. cohaerans*
	045553	*Gliocladium virens*			*P. daccinhae*
	104624	*Penicillium funiculosum*			*P. delafeldii*
	286366	*Trichoderma hazzianum*		11753	*P. species*
C3	9335	*Desufovibrio desufaricans*	C8		*Pseudomonas fuorescens*
	8307				*P. deva*
	8313				*P. fragi*
	9332	*Desufovibrio gigas*			*P. haithinsis*
	8303	*Desufovibrio vulgaris*			*P. lenoignei*
	8406				*P. ochracrea*
	8446				*P. pictorum*
	8457				*P. saccharophila*
					P. salopia
C4	9890	*Flavobacterium capsulatum*		23483	*P. species ATCC (Mn)*
	11299	*F femigineum*		11751	*P. species*
	8767	*F. resinovonum*			
	10741	*F. species*	C9		*Pseudomonas putida*
	12722	*F. species*			*Pseudomonas species*
	9491	unnamed *Flavobacterium*			
	10395		C10	8345	*Thiobacillus concretivorus*
	11635			9548	*T. denitrificans*
				9490	*T. ferroxidans*
C5	11805	*Mycobacterium album*		8539	*T. neopolitanus*
	9738	*M. favum*		8370	*T. thioparus*
	10420	*M. parafinicum*			
	11807	*M vaccae*	C11	Isolates from reported cases of sealant degradation	
C6	11222	*Nocardia amanae*			
	9438	*N. petroleophila*			
	11399	*Nocardia species*			
	12811				
	12814				

* The numerical annotation is the International Species Reference Number.

Cultures were obtained from the following sources:

- The National Collections of Industrial and Marine Bacteria (NCIB)
  23 St Machar Drive
  Aberdeen
  Scotland AB2 1RY

- American Type Culture Collection (ATCC)
  Rockville
  Maryland
  USA

Isolates were obtained from sites where microbial degradation of sealants was apparent. Selection of cultures was made on the basis of direct or inferred ability to undertake biodegradation of polymeric materials.

A3.2 ENVIRONMENTAL CONDITIONS

The following environmental sites were used for challenge testing. From the data obtained, correlation may be drawn with specific consortia (Section A3.3).

E1 Sewage outfall (*Sew out*).
This site was the discharge effluent stream from the Severn Trent Water Finham Waste Treatment Works. Samples were fully submerged in this aerobic environment.

E2 Sewage aerobic/99 per cent humidity (*99% humidity*).
Output stream from sludge digester at the Severn Trent Water Finham Waste Treatment Works. Samples were suspended in a closed environment in which the humidity approached 99 per cent.

E3 Sewage anaerobic (*Sew anaer*).
Samples were placed in an activated sludge tank.

E4 Raw water (upland) (*Yorks pre*).
Samples were placed in the incoming, untreated, raw water to the Albert Water Treatment Works, Yorkshire Water. The characteristics of this water are given in table A3.1.

E5 Upland treated water (*Yorks post*).
Samples were suspended in Stansfield View Service Reservoir. The characteristics of this water are given in Table A3.2.

Table A3.2 *Characteristics of raw water to Albert Treatment Works and upland treated water of Stansfield Service Reservoir*

	Raw water (upland) Albert Treatment Works	Upland water (treated) Stansfield Service Reservoir
pH	5.15	8.02
Conductivity	79.07	113.56
Turbidity	8.85	0.23
Colour	45.19	3.26
Nitrogen (ammoniacal)	0.08	
Aluminium	0.93	0.075
Manganese	0.13	0.008
Iron	0.95	0.04
Chlorine-free		0.077
Chlorine total		0.104

E6 Raw water (lowland river) (*Tewkes pre*).
    Samples were suspended in River Severn water entering the Severn Trent Water Tewkesbury Water Treatment Works.

E7 Lowland treated water (*Tewkes post*).
    Samples were suspended in the final outflow from the Severn Trent Water Tewkesbury Water Treatment Works.

E8 Samples were buried in soil (Cryfield, University of Warwick) and stored at 21 °C (*Soil*).

    Dry controls (*Dry*).

    Wet controls. Sterilised Swithland reservoir water (*Wet*).

Test strips of each sealant type, three for each time-point (ie three and six months), were threaded onto nylon and suspended in wire baskets. Each basket was immersed in 3 ppm hypochlorite for 10 minutes before placement in the challenge environment.

A3.3 CORRELATION BETWEEN ENVIRONMENTAL CONDITIONS AND CONSORTIA

From the challenge testing undertaken it is possible to draw correlations between the effect particular consortia had on sealant formulations with that produced in specified environmental locations. These correlations are summarised in Table A 3.3.

Table A3.3 *Correlation between environmental challenge and specific consortia as judged by the effect on sealant performance and surface topography*

Environment		Significant consortia
Sewage outfall	(sew out)	C1, C2, C3, C5, C6, C7, C8, C9, C10
Sewage anaerobic/99 per cent humidity	(99% humidity)	C1, C2, C3, C5, C6, C7, C8, C9, C10
Sewage anaerobic	(sew anaer)	C3, C7, C8, C9, C10
Raw water (upland)	(Yorks pre)	C4, C5, C6, C7, C8, C9, C10
Upland treated water	(Yorks post)	C4, C7, C8, C9, C10
Raw water (lowland river)	(Tewkes pre)	C4, C5, C6, C7, C8, C9, C10
Lowland treated water	(Tewkes post)	C4, C7, C8, C9, C10
Soil	(Soil)	C1, C2, C5, C6, C7, C8, C9

Caution must be exercised in utilising these comparisons.

The diversity of micro-organisms in the chosen environments was known to be such that virtually every species used in the consortia challenges will be present. The situation is further complicated by the sealant formulation *per se*, ie some components are susceptible to microbial attack while others are less so.

In selecting consortia, consideration must therefore be given to both the microbial species and the polymer composition.

From the above, the summary data sheets (Appendix A4) and Tables 7.1, 7.2, 7.3, 7.4, 7.5, 7.6, 7.7 and 7.8, a substantial database has been established against which an informed choice can be made about the consortia that have the greatest potential to compromise a particular sealant formulation.

A4 Summary data sheets: sealant strips

Key

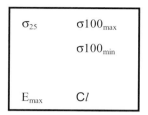

σ_{25}	Mean value of σ at 25 per cent extension in replicate strip tensile tests.
$\sigma_{100}(max)$	Maximum value of stress at 100 per cent extension in strip tensile tests (MPa).
$\sigma_{100}(min)$	Minimum value of stress at 100 per cent extension in strip tensile tests (MPa).
E_{max}	Mean E value at maximum force readings in tensile extension tests.
Δl	Mean change in length of replicate strip test specimens after testing (mm).
0	No change in length measurement.
F	No sample available.

All samples, except the dry control, were tested wet.

Table A4.1 Polyurethane – consortia – summary of first time-point mean results *(three months' exposure)*

	Dry		Wet		C1		C2		C3		C4		C5		C6		C7		C8		C9		C10		C11	
t1	0.28	1.178 1.138	0.35	1.525 1.355	0.85	1.754 1.561	0.7	2.082 1.587	0.73	3.456 2.342	0.05	0.1913 0.1265	0.05	0.2027 0.191	0.07	0.2954 0.2669	0.05	0.1973 0.1329	0.05	0.2066 0.1786	0.05	0.192 0.143	0.05	0.2069 0.1882	0.06	0.2539 0.2346
	140.4	0	140.5	0	73.58	0	93.58	0	111	0.13	116	0	147	0	144.9	0	134.5	0	130.8	0	100.2	0.1	140.6	0	152.9	0
t2	0.42	1.072 0.9169	0.51	1.195 0.4737	0.6	2.112 0.998	0.11	1.36 0.9146	1.65	4.314 2.332	0.08	0.222 0.1493	0.1	0.2244 0.1493	0.1	0.2205 0.1589	0.09	0.2412 0.1365	0.1	0.2084 0.1767	0.09	0.2161 0.1785	0.11	0.2345 0.1933	0.1	0.2286 0.09217
	61.8	0	56.52	0	24.23	0	21.25	0	58.85	0	66.12	0	56.18	0.1	50.13	0	44.12	-0.03	49.38	0	52.15	0	51.65	0	54.93	0
t3	0.35	1.395 0.7824	0.29	0.9583 0.5217	0.67	2.058 0.5021	0.86	2.085 1.143	1	2.806 1.312	0.07	0.1996 0.1338	0.07	0.2365 0.1946	0.05	0.1451 0.1338	0.07	0.24 0.2124	0.08	0.2173 0.1947	0.07	0.2208 0.159	0.07	0.7454 0.2308	0.07	0.1952 0.1582
	53.25	0	84.73	0	46.82	0	58.02	0	45.93	0	68.23	0.1	81.98	0	75.47	0	86.03	0.03	74.75	0	78.97	0	93.12	0	66.17	0
t4	0.27	1.138 0.8699	0.32	1.074 0.7018	0.63	2.01 0.9605	0.89	2.636 1.507	1.03	3.542 3.295	0.72	2.28 0.217	0.08	0.2218 0.1998	0.07	0.1952 0.1876	0.05	0.1458 0.1336	0.07	0.2218 0.2059	0.07	0.1963 0.1836	0.11	0.3812 0.2731	0.07	0.2009 0.04443
	84.93	0	69.93	0	55.77	0	89.03	0	80.35	0	79.33	0	69.82	0	71.1	0	65.62	0	76.88	0.1	75.43	0.1	78.03	0.1	73.95	0
t5	0.16	0.7472 0.4095	0.46	1.174 1.023	1.06	1.978 0.8764	1.29	3.402 2.377	2.15	5.736 4.951	0.1	0.2505 0.215	0.11	0.2803 0.2553	0.15	0.3471 0.1413	0.13	0.2918 0.2458	0.02	0.1412 0.1179	0.03	0.01319 0.1149	0.03	0.1239 0.09957	0.14	0.3577 0.2905
	79.55	0	65.17	0	43.53	0	71.1	0	67.16	0	69.7	0	74.17	0	68.02	0	61.88	0.03	82.37	0.13	76.54	0	73.57	0	68.87	0
t6	0.21	0.7488 0.1467	0.23	0.4082 0.179	0.15	0.5833 0.06112	0.67	1.91 1.454	0.84	2.186 1.896	0.05	0.1056 0.03169	0.05	0.1357 0.08269	0.06	0.1951 0.06359	0.06	0.1365 0.126	0.05	0.1592 0.1287	0.05	0.1236 0.03991	0.05	0.07897 0.01861	0.05	0.1348 0.08371
	79.15	0	41.63	0	20	0	87.2	-0.17	62.68	-0.23	43.08	-0.1	62.75	-0.13	72.73	-0.2	72.88	-0.03	84.82	-0.2	57.15	0.13	30.88	-0.23	59.48	-0.2
t7	0.17	0.4475 0.2773	0.19	0.4966 0.2139	0.06	0.3037 0.1136	0.53	1.323 1.052	0.65	1.692 0.924	0.04	0.07539 0.04933	0.04	0.107 0.07586	0.04	0.1259 0.1099	0.05	0.106 0.09449	0.04	0.1258 0.09685	0.04	0.09377 0.07641	0.04	0.09126 0.08244	0.04	0.07847 0.05375
	57.55	0	28.83	0	32.45	0	62.52	0	49.48	-0.1	42.5	0	52.72	0	73.78	0	57.4	0	68.75	0	54.12	-0.03	50.3	-0.03	45.92	0
t8	0.4	1.18 0.8068	0.32	1.713 1.254	0.65	3.445 1.364	1.52	6.017 5.348	0.82	3.419 2.212	0.06	0.3709 0.1967	0.07	0.1869 0.1714	0.05	0.6694 0.05689	0.1	0.6694 0.4078	0.11	0.5333 0.3642	0.1	0.4517 0.2647	0.1	0.2803 0.2188	0.11	0.2404 0.2114
	52.68	0	96.38	0	71.12	0	94.68	0.4	77.18	0.47	81.37	1.17	59.02	0	56.25	0.3	111.8	0.17	103.8	0.33	82.18	-0.2	66.05	0.4	58.38	-0.06
t9	0.12	0.3345 0	0.11	0.1859 0.1439	0.14	0.4154 0.01455	0.39	0.6393 0.5496	0.18	0.2586 0.1698	0.03	0.04685 0.02875	0.02	0.03143 0.0236	0.02	0.05582 0.01509	0.05	0.04752 0.03955	0.04	0.08732 0.07547	0.03	0.0585 0.05094	0.02	0.04734 0.01483	0.04	0.01034 0.02872
	77.8	0	36.42	0	43.1	0	43.1	-0.1	31.27	0	36.57	-0.07	28.32	0	38.25	-0.33	117.5	0	61.52	-0.03	45	-0.1	30.43	-0.1	34.35	0

Table A4.2 Polyurethane – consortia – summary of second time-point mean results (six months' exposure)

	Dry		Wet		C1		C2		C3		C4		C5		C6		C7		C8		C9		C10		C11	
PU1	0.39	1.401 1.146	0.57	1.517 1.285	0.21	0.5681 0.3123	0.2	0.3981 0.02145	0.38	1.117 0.8877	0.82	0.9439 0.7837	0.26	0.841 0.5982	0.44	1.379 1.135	0.38	0.7788 0.6587	0.31	0.9144 0.07259	0.36	0.8676 0.7352	0.28	0.9423 0.7151	0.37	1.245 0.8736
	111.28	0.00	87.02	0.00	55.05	0.13	52.05	0.20	86.70	0.07	77.13	0.10	94.58	0.10	113.42	0.00	64.62	0.00	93.37	0.00	84.35	0.00	93.30	0.00	111.18	0.00
PU2	0.62	1.313 0.9092	0.87	1.065 0.71154	0.76	1.187 0.8611	0.6	0.8437 0.3933	0.71	0.9986 0.4164	0.78	0.8239 0.5408	0.02	0.4328 0.2793	0.72	1.141 0.6182	0	0.523 0.2989	0.61	1.153 0.9472	0.2	0.7063 0.3675	0.75	0.871 0.5465	0.7	0.9889 0.6315
	46.87	-0.10	28.15	0.10	36.13	0.07	30.28	0.10	36.90	-0.03	21.02	0.07	20.38	0.10	45.00	0.00	16.12	-0.03	45.90	0.07	33.08	0.00	24.60	0.00	36.40	0.07
PU3	0.66	1.526 1.26	0.56	0.8846 0.7833	0.36	0.9133 0.5016	0.43	0.4831 0.3397	0.44	0.9674 0.7763	0.6	1.007 0.555	0.39	1.1128 0.8482	0.5	1.206 0.7702	0.47	0.8238 0.4284	0.43	1 0.5471	0.37	0.6794 0.3663	0.43	0.7543 0.4364	0.54	0.8666 0.718
	58.28	-0.10	39.55	-0.03	64.78	0.07	26.60	0.10	50.90	0.10	28.20	0.17	64.98	0.03	59.23	0.03	44.15	0.00	47.05	0.03	33.08	0.00	24.50	0.00	37.55	0.00
PU4	0.54	1.21 0.8772	0.57	1.305 0.8649	0.38	1.073 0.6843	0.44	0.9852 0.9407	0.52	1.267 1.121	0.647	1.221 0.9663	0.47	1.055 0.6855	0.55	1.014 0.8126	0.55	1.335 0.7654	0.41	0.8578 0.7131	0.46	1.251 0.6856	0.54	0.9786 0.7627	0.47	0.9193 0.4745
	47.25	0.00	47.83	0.10	65.78	0.13	54.60	0.13	63.37	0.03	43.58	0.07	45.28	0.00	44.77	0.00	63.48	0.00	53.30	0.03	48.30	0.03	44.37	0.07	48.30	0.03
PU5	0.72	1.417 0.794	0.75	1.483 1.178	0.59	1.386 0.7041	0.7	1.55 0.9554	0.68	1.65 0.8965	0.73	1.007 0.5522	0.66	1.67 1.289	0.77	1.519 0.8781	0.81	1.157 1.081	0.16	0.6131 0.4381	0.24	0.6521 0.3762	0.23	0.553 0.2097	0.88	1.727 0.9653
	51.68	-0.03	51.55	0.00	66.90	0.00	56.15	0.07	66.65	0.07	34.83	0.10	62.52	0.00	44.33	0.00	35.90	0.00	64.22	0.00	49.22	0.10	54.10	0.07	41.10	0.00
PU6	0.4	0.8419 0.575	0.37	0.7025 0.4034	0.27	0.6494 0.5819	0.22	0.7271 0.2386	0.26	0.5751 0.4308	0.39	0.9455 0.5819	0.21	0.5174 0.3647	0.27	0.7587 0.5156	0.27	0.4595 0.0829	0.26	0.4658 0.2415	0.21	0.3155 0.2779	0	0.3102 0	0.35	0.5807 0.1811
	54.28	-0.10	45.40	-0.10	62.22	-0.20	73.70	-0.20	54.20	-0.20	62.45	-0.13	54.70	-0.10	72.05	-0.20	34.93	-0.20	42.20	-0.20	34.90	-0.10	19.15	-0.30	43.28	-0.27
PU7	0.24	0.4927 0.4417	0.33	0.469 0.3413	0.18	0.3161 0.1615	0.24	0.3665 0.1794	0.19	0.5783 0.2137	0.31	0.3719 0.2567	0.22	0.5639 0.1348	0.24	0.6112 0.3056	0.28	0.4065 0.2381	0.28	0.5023 0.2467	0.2	0.6142 0.1887	0.3	0.321 0.2697	0.27	0.3098 0.1402
	49.43	0.12	30.68	-0.10	29.25	0.10	27.82	0.00	39.83	-0.03	24.93	0.03	38.70	0.40	48.72	-0.03	28.28	0.03	33.30	0.00	33.47	0.00	25.70	0.00	22.27	0.10
PU8	0.58	1.597 1.038	0.6	1.926 1.446	0.51	1.866 1.332	0.45	1.756 1.39	0.46	1.863 0.9262	0.62	1.99 1.38	0.33	0.8538 0.63	0.61	1.482 0.648	0.69	2.572 1.203	0.56	2.445 1.5	0.73	3.071 2.767	0.7	3.308 1.603	0.79	2.101 1.359
	45.13	0.06	60.38	0.57	67.00	0.60	77.53	0.60	65.48	0.10	64.60	0.60	54.37	0.40	45.23	0.27	66.70	0.10	80.70	0.20	89.12	0.00	82.97	0.00	55.43	0.40
PU9	0.17	0.2958 0.208	0.13	0.2611 0.2504	0.11	0.2312 0.1277	0.06	0.1217 0.07568	0.09	0.1492 0.1129	0.12	0.3079 0.1448	F	- -	0.09	0.1487 0.1201	0.11	0.1735 0.1404	0.19	0.4239 0.2064	0.18	0.2314 0.1761	0	0.05379 0.008335	0.18	0.2776 0.09383
	40.08	0.03	54.23	-0.13	38.85	-0.13	48.78	-0.23	38.12	-0.03	57.13	0.00	-	-	40.43	-0.30	36.10	-0.13	42.72	-0.07	30.03	0.00	19.62	-0.13	25.10	0.00

Table A4.3 *Polyurethane – environmental – summary of first time-point mean results (three months' exposure)*

	Dry		Wet		99% Humidity		Sewage out		Sewage an.		York pre.		York post.		Tewks pre		Tewks post		Soil	
PU 1	0.28	1.178	0.35	1.525	0.32	0.7219	0.32	0.8713	0.29	0.7086	0.20	0.3908	0.33	0.9327	0.21	0.6521	0.16	0.918	0.43	0.8083
	140.4	1.138	140.5	1.355	83.31	0.5034	77.72	0.6124	73.90	0.5774	28.08	0.04289	81.53	0.7424	74.98	0.4718	115.03	0.4986	45.47	0.5979
		0		0		0.03		0.17		0		0.1		0.03		0.2		0.33		-0.23
PU 2	0.42	1.072	0.51	1.195	0.52	1.178	0.58	0.5929	0.59	1.064	0.43	1.047	0.50	0.671	0.48	1.042	0.17	0.2101	0.47	1.038
	61.8	0.9169	56.52	0.4737	23.65	0.3374	22.50	0.3388	49.53	0.427	39.42	0.4634	24.35	0.2358	39.05	0.5756	28.40	0.08529	48.38	0.4754
		0		0		-0.03		0.07		0		-0.03		0		0.2		0.37		0.03
PU 3	0.35	1.395	0.29	0.9583	0.50	1.028	0.43	0.8706	0.44	1.19	0.43	1.448	0.39	0.6101	0.42	0.7958	0.21	0.4323	0.48	1.244
	53.25	0.7824	84.73	0.5217	43.50	0.5821	41.75	0.5436	52.62	0.5535	54.07	0.4608	37.07	0.4892	39.27	0.5479	43.90	0.2911	72.80	1.19
		0		0		-0.1		0.1		0		-0.1		-0.07		0.07		0.2		-0.1
PU 4	0.27	1.138	0.32	1.074	0.37	0.9062	0.40	0.7422	0.38	0.7563	0.30	0.7286	0.32	0.9313	0.34	0.6198	0.17	0.7925	0.38	1.154
	84.93	0.8699	69.93	0.7018	41.97	0.2571	38.03	0.3753	42.03	0.4526	62.53	0.2139	34.70	0.4243	28.48	0.2551	40.88	0.2657	72.47	0.9323
		0		0		0.07		0.17		0		-0.03		0.17		0.6		0.77		0
PU 5	0.16	0.7472	0.46	1.174	0.56	0.8104	0.55	0.8742	0.63	1.068	0.46	1.357	0.56	1.07	0.61	0.9399	0.18	0.2699	0.52	1.386
	79.55	0.4095	65.17	1.023	28.03	0.4149	30.77	0.5021	35.60	0.7232	47.15	0.3313	37.55	0.5452	28.82	0.5029	30.57	0.179	63.15	1.037
		0		0		-0.1		-0.03		0.1		-0.13		-0.03		0.17		0.33		-0.13
PU 6	0.21	0.7488	0.23	0.4082	0.41	1.179	0.44	0.5532	0.40	0.6244	0.33	0.9328	0.30	0.3793	0.39	0.5304	0.06	0.09113	0.26	0.7822
	79.15	0.1467	41.63	0.179	21.30	0.2908	26.70	0.2933	48.48	0.3044	66.18	0.4228	24.30	0.1095	35.83	0.3757	27.83	0.05129	58.00	0.2935
		0		0		-0.93		-0.87		2.3		-1		-0.1		0.37		1		-0.13
PU 7	0.17	0.4475	0.19	0.4966	0.20	0.3501	0.30	0.4805	0.25	0.3579	0.24	0.3911	0.00	0.3535	0.00	0.1652	0.08	0.2352	0.18	0.4598
	57.55	0.2773	28.83	0.2139	36.40	0.2312	33.95	0.2674	36.78	0.168	34.20	0.1268	15.15	0.1119	12.33	0.04599	39.33	0.08798	57.95	0.3387
		0		0		0		0		0		-0.07		0.1		0.4		0.8		0
PU 8	0.4	1.18	0.32	1.713	0.50	1.08	0.58	0.8763	0.39	0.5348	0.43	0.6944	0.40	0.607	0.44	0.5547	0.25	0.6062	0.47	1.172
	52.68	0.8068	96.38	1.254	41.10	0.1013	27.90	0.4701	29.48	0.3359	35.90	0.5251	30.25	0.3806	27.57	0.4687	33.12	0.2062	55.78	0.8068
		0		0		0.07		0.47		0.38		0.13		0.2		0.4		0.37		-0.03
PU 9	0.12	0.3345	0.11	0.1859	0.18	0.2888	0.04	0	0.17	0.3038	0.09	0.1645	0.20	0.3758	0.00	0.3758	0.01	0.05994	0.18	0.3805
	77.8	0	36.42	0.1439	35.88	0	21.98	0	37.26	0.2134	35.73	0.1025	39.38	0.2547	12.88	0.2547	68.20	0	31.38	0.2771
		0		0		-0.05		0.05		0		-0.17		-0.03		1.1		2.35		-0.3

Table A4.4 *Polyurethane – environmental – summary of second time-point mean results (six months' exposure)*

	Dry		Wet		99% Humidity		Sewage out		Sewage an.		York pre.		York post.		Tewks pre		Tewks post		Soil	
PU1	0.39	1.401 1.146	0.57	1.517 1.285	0.44	1.151 0.5744	0.42	1.022 0.7186	0.51	1.174 0.7484	0.42	0.6253 0.4648	0.51	1.405 0.8043	0.32	0.7628 0.2907	0.18	0.8372 0.5283	0.43	0.5833 0.4006
	111.28	0.00	87.02	0.00	65.55	0.00	59.48	0.17	72.58	0.00	46.15	0.03	110.60	0.00	66.01	0.17	114.97	0.43	44.90	0.23
PU2	0.62	1.313 0.9092	0.87	1.065 0.7154	0.79	1.177 0.573	0.65	1.117 0.6602	0.00	0.7712 0.5538	0.74	1.097 0.5374	0.82	0.967 0.7478	0.78	1.164 0.4631	0.31	0.6901 0.195	0.70	1.142 0.8062
	46.87	-0.10	28.15	0.10	39.30	0.00	34.65	0.10	18.50	0.20	16.83	0.00	25.37	0.00	22.45	-0.03	22.73	0.37	36.77	0.00
PU3	0.66	1.526 1.26	0.56	0.8846 0.7833	0.72	1.252 0.7788	0.44	0.6488 0.2166	0.74	0.7421 0.4288	0.68	1.495 0.9706	0.61	1.295 0.6081	0.66	1.021 0.675	0.25	0.7591 0.5698	0.63	1.259 0.9284
	58.28	-0.10	39.55	-0.03	29.48	0.07	33.28	0.00	20.07	0.13	52.95	0.00	60.48	0.00	33.00	-0.07	66.47	0.30	50.48	-0.10
PU4	0.54	1.21 0.8772	0.57	1.305 0.8649	0.63	1.119 0.7796	0.53	0.9827 0.2256	0.69	0.9295 0.641	0.59	1.106 0.5125	0.58	1.258 0.797	0.45	0.7384 0.5317	0.18	0.6672 0.3852	0.48	0.8879 0.8289
	47.25	0.10	47.83	0.10	31.43	0.10	18.80	0.00	30.05	0.20	36.15	0.10	47.22	0.17	34.13	0.23	67.32	0.87	47.27	0.03
PU5	0.72	1.417 0.794	0.75	1.483 1.178	0.79	1.073 0.6019	0.58	1.083 0.5523	0.93	1.601 0.9275	0.75	1.375 0.5546	0.84	1.395 1.149	0.68	0.88387 0.6949	0.22	0.3113 0.1988	0.69	1.079 0.9442
	51.68	-0.03	51.55	0.00	31.62	-0.07	29.90	0.00	36.97	0.17	20.05	-0.10	43.25	0.00	29.05	-0.10	27.12	0.30	41.07	0.00
PU6	0.40	0.8419 0.575	0.37	0.7025 0.4034	0.56	1.306 0.9169	0.47	0.6099 0.4395	0.64	1.529 1.386	0.49	0.6146 0.4385	0.39	1.015 0.2084	0.39	1.108 0.3992	0.07	0.09194 0.07247	0.35	0.9665 0.4565
	54.28	-0.10	45.40	-0.10	70.98	-1.03	26.05	-1.30	91.73	-1.15	24.53	-0.90	61.53	-0.20	59.70	-1.07	55.15	0.95	39.73	-0.23
PU7	0.24	0.4927 0.4417	0.33	0.469 0.3413	0.29	0.6293 0.3837	0.15	0.2704 0.1611	0.43	0.4352 0.1488	0.22	0.2503 0.1913	0.23	0.4275 0.2606	0.16	0.284 0.1096	0.12	0.356 0.1155	0.20	0.4754 0.2353
	49.43	0.07	30.68	0.00	20.86	0.00	22.73	0.00	18.68	0.03	20.08	0.20	39.33	0.10	32.68	-0.07	46.15	0.57	41.03	0.03
PU8	0.58	1.597 1.038	0.60	1.926 1.446	0.89	2.158 1.971	0.49	0.7561 0.3917	0.70	1.731 1.072	0.63	1.089 0.4782	0.56	1.674 1.195	0.55	0.8757 0.5149	0.27	0.6886 0.1834	0.53	0.9112 0.7135
	45.13	-0.03	60.38	0.57	53.60	0.40	34.90	0.37	49.02	0.00	34.68	0.20	54.18	0.10	27.78	0.30	42.50	0.33	39.63	0.33
PU9	0.17	0.2958 0.208	0.13	0.2611 0.2504	0.08	0.08034 0	0.09	0.1442 0.08441	0.19	0.3083 0.1947	0.12	0.1765 0.1178	0.16	0.2731 0.09551	0.00	0.04075	0.00	0.06528 0	0.14	0.2385 0.1383
	40.08	0.03	54.23	0.03	15.20	-0.10	32.58	-0.10	27.88	0.00	32.32	-0.10	28.85	-0.10	12.35	-0.10	10.20	1.50	24.85	-0.40

Table A4.5 Silicone/polysulfide – consortia – summary of first time-point mean results (three months' exposure)

	Dry		Wet		C1		C2		C3		C4		C5		C6		C7		C8		C9		C10		C11	
Si 1	0.13	1.297 1.016	0.05	0.8724 0.07959	0.14	2.855 2.654	0.31	2.465 1.533	0.32	2.161 1.285	0.03	0.2532 0.204	0	0.1907 0.1496	0	0.2542 0.1938	0	0.2518 0.06766	0.01	0.2195 0.1546	0.01	0.3147 0.2342	0	0.28 0.1208	0	0.2336 0.05634
	474.45	0	416.1	0	431.7	0	436.5	0	310.2	0	428.1	-0.13	56.24	0	32.21	0	223.05	0.03	48.16	0.03	28.57	0.17	95.61	0.13	208.74	0.1
Si 2	0.1	0.608 0.5103	0.08	0.5159 0.3307	1.15	3.002 1.479	0.35	1.251 1.144	0.11	1.882 0.5709	0.03	0.1335 0.1187	0	0.252 0.1875	0	0.2063 0.1293	0.06	0.2042 0.1112	0.03	0.1813 0.1874	0.05	0.2181 0.1838	0.04	0.2204 0.19	0.04	0.2122 0.1808
	284.9	0	208.25	0	211.4	0	194.4	0	307.6	0	199.1	0.03	88.86	0	254.97	0	184.02	0	247.97	0	268.23	0.13	317.03	0	244.97	0
Si 3	0.03	0.9566 0.8924	0.03	0.8038 0.6006	1.2	2.995 2.674	0	1.668 1.436	0.17	2.872 2.335	0.01	0.2209 0.198	0.01	0.2001 0.158	0.01	0.2344 0.1571	0.01	0.2928 0.2493	0.23	0.3089 0.2181	0.01	0.3443 0.2479	0.03	0.3098 0.2487	0.03	0.3086 0.257
	762	0	496.9	0	468	0	427.5	0	414.8	0	560.6	0	458.8	0	408.33	0	460.13	0.07	432.43	0	490.1	0	455.23	0	436.37	0
PS 1	0.05	0.489 0.4782	0	0.01397 0.1108	0.52	1.214 0.9042	0.12	0.8837 0.8391	0	0.5695 0.4975	0	0.0882 0.04305	0.05	0.1998 0.07545	0	0.06006 0.05493	0	0.08371 0.07834	0	0.08468 0.08361	0	0.1045 0.07478	0.02	0.08504 0.07159	0.02	0.1041 0.08436
	85	0	352.4	2.13	752.7	0	716.7	0.93	573.37	0.5	810.4	0.6	882.93	0.6	592.7	0.67	847.3	0.5	781.2	0.43	646.2	0.8	409.07	1	632.2	0.4
PS 2	0.1	0.6516 0.5003	0.01	0.2371 0.1886	0.3	0.8554 0.6564	0.21	0.9793 0.7528	0.1	1.428 0.9942	0	0.0822 0.06029	0	0.09429 0.07222	0	0.07685 0.05516	0	0.06968 0.03962	0.01	0.1222 0.1075	0	0.087 0.07366	0.01	0.08059 0.06578	0	0.08884 0.06457
	652	0	523.6	0.73	446	0	639.8	0.63	618.8	0.27	907.5	0.23	807.57	0.6	622.57	0.67	835.67	0.2	874.03	0.27	865	0.23	751.83	0.83	819.67	0.3
PS 3	0.05	0.5301 0.4347	0.02	0.254 0.1518	0.2	1.098 0.8254	0.13	1.031 0.7928	0.04	1.25 0.7364	0	0.09579 0.08325	0	0.0903 0.0587	0	0.09605 0.07587	0	0.1257 0.08131	0.01	0.1139 0.09533	0	0.0809 0.06631	0.01	0.07641 0.06769	0	1.25 0.7364
	737.8	0	634.4	0.87	630.6	0	518.4	0.23	443.8	0.3	899.5	0.3	542	0.6	524.03	0.73	698.8	0.2	625.87	0.27	607.47	0.3	612.3	0.63	722.1	0.37
PS 4	0.06	0.6051 0.3731	0	0.1682 0.1248	0.13	0.9839 0.7641	0.19	1.442 0.7739	0	1.065 0.835	0	0.11007 0.09345	0	0.09788 0.08262	0	0.08326 0.07019	0.01	0.0838 0.04543	0.01	0.06611 0.0593	0.01	0.1113 0.1058	0.01	0.04575 0.01838	0.01	0.06233 0.04406
	595.5	0	504.5	1.6	688.9	0	620.8	0.73	709.4	0.27	909.5	0.5	785.77	0.5	541.2	0.93	624.63	0.5	730.63	0.5	696.9	1	307.33	1.6	687.3	0.67
PS 5	0.11	0.3056 0.2575	0.05	0.1794 0.06686	0.47	0.7753 0.5055	0.26	1.076 1.026	0.59	1.079 0.557	0.07	0.3744 0.03236	0.03	0.0698 0.03236	0.04	0.07031 0.0579	0.04	0.1175 0.06083	0.06	0.08506 0.06251	0.07	0.09551 0.08308	0.67	0.1384 0.06221	0.06	0.1113 0.09404
	67.65	0	37.08	0.77	35.58	0	90.88	0.53	35.33	0.13	83.23	0.5	42.57	0.47	52.6	0.5	52.42	0.53	28.63	0.3	33.35	0.33	49.88	0.4	57.32	0.67

Table A4.6 Silicone/polysulfide – consortia – summary of second time-point mean results (six months' exposure)

		Dry	Wet	C1	C2	C3	C4	C5	C6	C7	C8	C9	C10	C11
Si1		0.1	0.15	0.1	0.12	0.17	0.17	0.15	0.15	0.17	0.15	0.21	0.2	0.21
		1.11 / 0.7255	1.061 / 0.618	0.9293 / 0.7609	0.9415 / 0.6559	1.048 / 0.6763	1.186 / 0.9187	0.9051 / 0.7755	1.306 / 0.7442	0.9289 / 0.755	1.303 / 0.8555	1.094 / 0.3133	1.157 / 0.9655	0.9351 / 0.6399
		301.73	335.85	332.2	351.35	334.3	280.23	327	301.83	295.93	347.2	249.4	276.9	216.2
		0	0	0	-0.03	0	0	0	0	0.07	0.1	0	0	0.03
Si2		0.16	0.2	0.2	0.18	0.17	0.27	0.18	0.2	0.22	0.25	0.25	0.28	0.27
		0.9231 / 0.5307	0.5803 / 0.3829	0.8989 / 0.6551	0.7591 / 0.5532	0.8586 / 0.7338	0.7463 / 0.679	0.782 / 0.4923	0.8952 / 0.593	0.8222 / 0.631	0.8223 / 0.6746	0.6799 / 0.5362	0.7837 / 0.6374	0.7601 / 0.6433
		169.67	70.1	128.75	194.75	245.67	108	225.9	253.75	188.1	115.85	83.72	114.08	112.6
		-0.07	0.33	0.1	0.03	-0.03	0.03	0.1	0	0.07	0.1	0.1	0	0
Si3		0.07	0.09	0.11	0.09	0.07	0.1	0.08	0.1	0.15	0.16	0.14	0.16	0.12
		0.8872 / 0.7351	0.852 / 0.6915	1.345 / 0.6576	1.088 / 0.7483	0.8291 / 0.7618	1.019 / 0.6558	0.9784 / 0.942	0.8747 / 0.693	1.053 / 0.754	0.9263 / 0.7142	0.9571 / 0.8496	0.9474 / 0.7999	1.021 / 0.5937
		440.53	313.7	299.1	337.73	357.37	274.27	433.8	268	293.13	248.2	285.77	274.5	233.7
		0.03	0.17	0	0	-0.1	-0.07	-0.07	-0.1	-0.1	-0.07	0	0.07	-0.1
PS1		0.08	0.02	0.03	0.05	0.04	0.05	0.03	0.04	0.06	0.06	0.06	0.07	0.06
		0.6285 / 0.4386	0.1901 / 0.1337	0.3681 / 0.1938	0.3511 / 0.3056	0.4599 / 0.2613	0.3635 / 0.2082	0.2909 / 0.218	0.357 / 0	0.4337 / 0.3485	0.4826 / 0.332	0.4797 / 0.4317	0.3313 / 0.291	0.4309 / 0.235
		551.77	207.9	354.3	323.87	540.53	404.9	409.07	442.5	490.63	401.3	490.3	221.93	395.6
		0.13	3.33	0.67	0.57	0.27	0.67	0.5	0	0.03	0.43	0.1	0.13	0.03
PS2		0.14	0.03	0.04	0.05	0.06	0.06	0.03	0.03	0.09	0.06	0.07	0.07	0.07
		0.743 / 0.6714	0.2453 / 0.1797	0.4031 / 0.2447	0.4191 / 0.3308	0.395 / 0.3246	0.5247 / 0.3556	0.33 / 0.1457	0.27 / 0.2032	0.4695 / 0.4075	0.6078 / 0.3804	0.4231 / 0.2899	0.4399 / 0.3557	0.5253 / 0.4071
		562.03	462.9	614.75	463.65	418.87	792.5	499.95	334.5	511.95	708.3	675.65	353.1	624.67
		0.06	1.17	0.5	0.67	0.1	0.23	0.1	0.03	0.1	0.07	-0.1	0	0.1
PS3		0.07	0.02	0.04	0.02	0.04	0.05	0.01	0.03	0.05	0.08	0.07	0.04	0.05
		0.4949 / 0.4759	0.2433 / 0.1884	0.3945 / 0.2377	0.1142 / 0.04748	0.3619 / 0.2534	0.6755 / 0.4566	0.2712 / 0.1225	0.2727 / 0.2383	0.397 / 0.3577	0.5913 / 0.4509	0.5159 / 0.3803	0.3304 / 0.2952	0.4075 / 0.295
		545.83	419.47	321.03	340.17	329.9	461.4	505.45	311.03	419.2	498.7	446.67	309.93	469.43
		0.06	1.77	0.17	0.1	0	0.4	0.07	0.03	-0.07	0.1	0	0.1	0.1
PS4		0.08	0.03	0.04	0.04	0.03	0.04	0.06	0.08	0.06	0.07	0.05	0.08	0.03
		0.6093 / 0.3926	0.2113 / 0.1226	0.3786 / 0.2476	0.3701 / 0.2231	0.2571 / 0.2015	0.3764 / 0.2513	0.3416 / 0.2955	0.5209 / 0.1444	0.4112 / 0.3289	0.4568 / 0.2286	0.3158 / 0.1939	0.3859 / 0.3142	0.2916 / 0.1771
		666.6	159.7	604.25	172.2	510.5	366.37	475.75	515.05	450.43	325.47	374.97	261.03	463.35
		0.13	2.57	0.6	0.67	0.13	0.4	0.33	0.1	-0.03	0.3	0.47	0.1	0.13
PS5		0.21	0.16	0.29	0.22	0.23	0.22	0.29	0.19	0.22	0.32	F	0.28	0.18
		0.4006 / 0.364	0.4433 / 0.09871	0.4735 / 0.3033	0.6235 / 0	0.4352 / 0.07696	0.7055 / 0.3294	0.5949 / 0.4045	0.4495 / 0.3921	0.4494 / 0.1945	0.5292 / 0.3411	- / -	0.5527 / 0.2407	0.5484 / 0.371
		63.55	112.7	41.4	81.18	54.83	112.85	71.75	67.83	41.23	36.43	-	37.78	86.97
		0.03	1.6	0.5	0.4	0.1	0.3	0.37	0.3	0.2	0.2	-	0.3	0.33

Table A4.7 *Silicone/polysulfide – environmental – summary of first time-point mean results (three months' exposure)*

	Dry		Wet		99% Humidity		Sewage out		Sewage an.		York pre.		York post.		Tewks pre		Tewks post		Soil	
Si 1	0.13	1.297	0.05	0.8724	0.1	0.6977	0.15	0.7047	0.12	0.5586	0.09	0.6974	0.13	0.688	0.12	0.8589	0.1	0.992	0.14	1.097
		1.016		0.07959		0.3495		0.3734		0.471		0.6112		0.5154		0.307		0.3729		0.5449
	474.45	0	416.1	0	160.7	0	204.63	0	258.73	0.03	275.53	0.03	249.43	-0.03	212.87	0.1	213.7	0.13	308.6	-0.13
Si 2	0.1	0.608	0.08	0.5159	0.14	0.5148	0.14	0.5521	0.14	0.4554	0.12	0.6286	0.11	0.4347	0.12	0.5146	0.05	0.2586	0.17	0.7991
		0.5103		0.3307		0.4668		0.4479		0.3785		0.3312		0.3176		0.3347		0.1365		0.5153
	284.9	0	208.25	0	113.07	0.13	143.73	0.2	178.1	-0.03	174.8	0.3	155.97	0.13	202.53	2.1	66.63	2.97	244.37	-0.17
Si 3	0.03	0.9566	0.03	0.8038	0.05	0.5951	0.06	0.6707	0.09	0.5301	0.04	0.7788	0.07	0.652	0.08	0.5356	0.04	0.596	0.06	1.13
		0.8924		0.6006		0.5021		0.5365		0.4661		0.5551		0.4089		0.4213		0.3111		0.8267
	762	0	496.9	0	316.5	-0.03	291.37	0.1	197.67	-0.1	413.9	0.13	211.5	0.03	178.43	0.8	546	1.27	401.4	-0.17
PS 1	0.05	0.489	0	0.1397	0.03	0.07068	0.3	0.1107	0.34	0.5677	0.04	0.2216	0.04	0.3322	0.04	0.221	0.02	0.134	0.11	0.6101
		0.4782		0.1108		0.05123		0.06374		0.3067		0.1629		0.1834		0.1223		0.1265		0.4306
	85	0	312.7	2.13	60.5	1	71.3	1.73	64.025	0.17	284.7	1.37	343.75	0.9	300	1.7	255.9	1.93	636.9	0.07
PS 2	0.1	0.6516	0.01	0.2371	0.03	0.05589	0.03	0.191	0.06	0.2425	0.11	0.3353	0.06	0.4083	0.04	0.2496	0.02	0.2453	0.2	0.5284
		0.5003		0.1886		0.01523		0.09621		0.1221		0.2523		0.2499		0.0775		0.1283		0.4821
	652	0	578	0.73	25.225	0.95	143.13	0.63	143.2	0.95	245	0.37	550.47	0.27	227.28	1.43	437.4	1.5	438.07	-0.2
PS 3	0.05	0.5301	0.02	0.254	0.19		0.04	0.0815	F	-	0.09	0.4444	0.02	0.05269	0.06	0.2828	F	-	0.21	0.9031
		0.4347		0.1518				0.01698				0.2693		0		0.1735				0.6225
	737.8	0	498.75	0.87	73.35	-0.9	20.1	-0.4	-	-	421.25	0.2	60.45	-	101.7	-	-	-	434.53	-0.57
PS 4	0.06	0.6051	0	0.1682	0.05	0.2026	0.02	0.06648	0.05	0.1933	0.04	0.2826	0.05	0.3919	0.04	0.1822	0.02	0.1446	0.13	0.6997
		0.3731		0.1248		0.1633		0.04674		0.1551		0.1784		0.3173		0.1354		0.109		0.3835
	692.8	0	461.41	1.6	236.9	1.9	63	1.67	192.07	2.6	260.65	1.27	662.67	0.97	158.3	1.83	252.63	2	608.33	-0.1
PS 5	0.11	0.3056	0.08	0.1794	0.24	0.3904	0.14	0.2098	F	-	0.17	0.3931	0.22	0.443	0.14	0.2184	0.16	0.2719	0.16	0.5448
		0.2575		0.06686		0.2093		0.1203				0.1124		0.2671		0		0.1164		0
	67.65	0	37.08	0.77	35.87	1.17	31.55	0.87	-	-	37.98	0.73	51.37	0.63	32.65	1.43	41.55	2.03	47.5	0

Table A4.8 Silicone/polysulfide – environmental – summary of first time-point mean results (six months' exposure)

		Dry		99% Humidity		Sewage out		Sewage an.		York pre.		York post.		Tewks pre		Tewks post		Soil	
Si 1		0.10	1.11	0.15	1.061	0.27	1.703	0.14	0.7469	0.21	0.9425	0.18	1.092	0.10	1.033	0.16	0.7326	0.14	0.7271	0.18	1.161
			0.7255		0.618		1.288		0.4248		0.5742		0.6685		0.581		0.4726		0.5713		0.6835
		301.73	0.00	335.85	0.00	342.66	−0.10	250.60	−0.10	276.05	0.00	236.90	0.00	276.87	−0.03	188.85	0.03	193.33	0.17	296.83	0.00
Si 2		0.16	0.9231	0.20	0.5803	0.24	0.8787	0.16	0.5212	0.24	0.7717	0.27	0.8653	0.26	0.8754	0.15	0.6772	0.08	0.7748	0.23	0.8624
			0.5307		0.3829		0.6899		0.4898		0.7036		0.6962		0.6375		0.4499		0.2301		0.5893
		169.97	−0.07	70.10	0.33	198.53	0.17	140.75	0.10	155.97	0.10	173.47	0.20	196.70	0.27	104.70	0.13	91.08	4.67	180.17	−0.13
Si 3		0.07	0.8872	0.09	0.852	0.14	0.9508	0.09	0.9405	0.18	1.341	0.13	0.9656	0.18	1.165	0.11	0.7851	0.02	0.4674	0.12	1.24
			0.7351		0.6915		0.6509		0.5974		0.7997		0.7946		0.8434		0.5671		0.221		1.114
		440.53	0.03	313.70	0.17	295.75	0.17	260.30	0.00	150.70	0.00	233.80	0.00	212.70	0.03	167.03	0.03	111.81	2.30	344.60	0.00
PS 1		0.08	0.6285	0.02	0.1901	0.07	0.5596	0.01	0.2253	0.03	0.1823	0.05	0.2745	0.04	0.2691	0.02	0.2873	0.01	0.08816	0.08	0.5394
			0.4386		0.1337		0.1036		0.1405		0.1237		0.2176		0.2458		0.1653		0.06032		0.4412
		551.77	0.13	207.90	3.33	117.05	2.53	250.35	3.00	129.40	2.85	216.43	1.93	310.00	1.73	208.07	3.00	108.30	2.73	515.67	0.43
PS 2		0.14	0.743	0.03	0.2453	F	−	0.02	0.242	F	−	0.23	0.3879	0.06	0.4977	0.03	0.1974	0.03	0.4258	0.26	0.4279
			0.6714		0.1797		−		0.03557		−		0.2434		0.3178		0.06946		0.1433		0.3625
		562.03	0.07	462.90	1.17	−	−	86.98	1.53	−	−	196.20	0.47	292.80	0.70	152.17	2.20	208.07	2.80	184.53	−0.07
PS 3		0.07	0.4949	0.02	0.2433	F	−	0.18	0.2043	F	−	0.11	0.3447	F	−	F	−	0.00	0.07641	0.33	0.8324
			0.4759		0.1884		−		0.1642		−		0.2654		−		−		−		0.5903
		545.83	0.23	419.47	1.77	−	−	26.82	−0.40	−	−	184.50	0.47	−	−	−	−	111.10	3.80	317.07	0.17
PS 4		0.08	0.6093	0.03	0.2113	F	−	0.03	0.3467	0.03	0.1363	0.04	0.3422	0.04	0.3127	0.03	0.3885	0.02	0.1775	0.16	0.514
			0.3926		0.1226		−		0.27021		0.1189		0.2459		0.1755		0.3056		0.04054		0.3371
		666.60	0.13	159.70	2.57	−	−	211.37	3.23	141.67	3.33	446.60	2.00	313.00	1.77	337.70	2.77	175.10	2.75	314.50	1.23
PS 5		0.21	0.4006	0.16	0.4433	0.18	0.2426	0.32	0.3953	0.42	0.577	0.24	0.5285	0.13	0.4075	0.36	0.6169	0.08	0.2936	0.28	0.4543
			0.364		0.09871		0.0566		0.3734		0.5568		0.5141		0.1497		0.5094		0.1262		0.3258
		63.55	0.03	112.70	1.60	26.30	0.90	38.18	1.57	43.63	0.07	98.35	1.00	27.65	1.07	62.18	0.63	36.03	3.37	50.83	1.33

CIRIA C520

115

A5 Summary data sheets: joints

F_{25}	Mean value of F at 25 per cent extension during replicate joint tensile extension tests.
F_{max}	Mean maximum value of F observed during replicate joint tensile extension tests.
E_{max}	Mean E value at maximum force readings in tensile extension tests.
F_{mode}	Mode of failure as defined below
TP1, TP2	First and second time-points for exposure
WIS rating	Water Industry Specification 4-60-01 March 1991: June 1

All samples, except the dry control, were tested wet. Mean results are presented (generally representing a minimum of three specimens). The bracketed results are merely informative and have been discounted in any further analysis of the results.

Modes of failure

Cohesive (C)	Failure in the bulk of the sealant
Adhesive (A)	Failure at the interface either between the sealant/primer or the primer/substrate. No sealant visible at the surface
Mixed (C/A)	Failure close to the interface. Some sealant clearly visible (mixed adhesive/cohesive failure) or sealant completely covering the surface (thin-film cohesive failure).

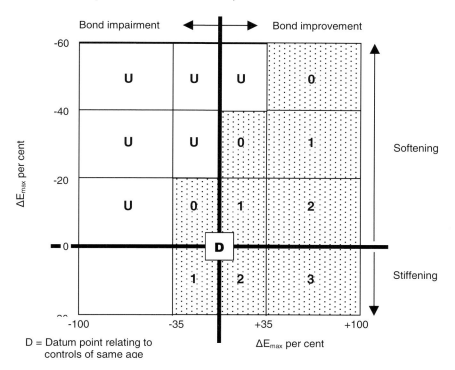

Figure A5.1 *Changes to modulus and extensibility after immersion (after WIS 4-60-01)*

In Figure A5.1, unshaded zones represent an *unacceptable* (U) change. Shaded zones represent a *potentially acceptable* change, the level of acceptability being indicated by the numbers 3 (most acceptable) to 0 (least acceptable). See section 3.4.2 for details.

Table A5.1 *Summary of first time-point mean results (three months' exposure)*

		Dry		Wet		Soil		Yorks pre		Yorks post		Tewkes pre		Tewkes post		99% humidity		Sewage outfall		Sewage anaerobic
PU3/Ep		423	681 (605)	410	646 (752)	400	696 (–)	250	409 (490)	463	683	337	446 (584)	400	614	326	403	350	508	No specimens
		43 (32)	Cx2 (C/Ax1)	54 (54)	C/Ax2 (Cx1)	56 (–)	Cx2 (Ax1)	46 (59)	Cx2 (C/Ax1)	50	C	39 (39)	Ax2 (Cx1)	44	C	35	A	36	A	
		427	707	373	663 (696)	347	464	355	377	385	401 (550)	336	382	380	420	325	313	323	328	No specimens
PU3/Isocy		43	C	51 (56)	Ax2 (C/Ax1)	36	A	27	C/A	27 (32)	Ax2 (Cx1)	27	Ax5	28	A	24	A	26	A	
PU6/Ep		No specimens received		No results		210	239 (293)	137	218 (159)	250	215 (329) (277)	F	160	240	200	F	131	F	145	No specimens
						31 (32)	Ax5 (C/Ax1)	40 (40)	Ax4 (C/Ax2)	22 (39) (26)	Ax4 (C/Ax1) (Cx1)	16	C/A	23	C/A	14	Ax6	18	Ax6	
PSl/Ep		83	292	68	214 (209)	63	252	47	212 (158)	70	249 (205)	72	193	63	206	57	172 (192)	60	186 (241)	No specimens
		528	C/A	411 (412)	Ax2 (C/Ax1)	565	C	453 (330)	Ax1 C/Ax1 (Cx1)	523 (380)	C/Ax2 (Ax1)	371	C/A	393	C/A	372 (384)	Ax2 (C/Ax1)	360 (523)	Ax2 (C/Ax1)	
PSl/Isocy		78	275 (164)	–	28	65	149 (170)	57	82	67	100	F	18	57	79	F	33	60	101	No specimens
		522 (180)	Cx2 (C/Ax1)	21	A	204 (352)	Ax2 (C/Ax1)	45	C/A	63	A	19	A	44	A	11	A	83	A	
PS3/Ep		No specimens received		No results		83	202 (204)	89	147	78	150	67	132	67	112	45	82	48	64	No specimens
						226 (224)	Ax6 (Cx1)	87	C/Ax6	109 (25–212)	Ax6	107	A	75	Ax6	63	Ax6	47	Ax6	
Si2/Acryl		223	455	200	373	203	399	200	412	190	336 (309)	180	362	213	394	263	405	207	406	No specimens
		156	C	193	C	177	C	214	C	153 (104)	Cx2 (C/Ax1)	143	C	163	C	172	C	174	C	

Key:
F_{25}
E_{max}
F_{max} Failure mode

u   unacceptable
0   least acceptable
3   most acceptable

A   failure at the sealant/primer interface
C   failure in the bulk of the sealant
C/A failure close to the interface

x   number of samples showing a particular mode of failure

Table A5.2 Summary of second time-point mean results (six months' exposure)

		Dry	Wet	Soil	Yorks pre	Yorks post	Tewkes pre	Tewkes post	99% humidity	Sewage outfall	Sewage anaerobic
PU3/Ep		493 / 819	400 / 642 (461)	380 / 533	430 / 570	367 / 661 (645)	395 / 535	383 / 540 (380)	F / 286	350 / 354	353 / 519 (448) (265)
		45 / C	45 (32) / Cx2 (Ax1)	35 / A	39 / CX2	49 (36) / Cx2 (Ax1)	40 / Cx2	37 (28) / Cx2 (C/Ax1)	22 / C/A	21 / A	38 (28) (25) / Cx4 (Ax1) (C/Ax1)
PU3/Isocy		417 / 562	380 / 517	F / 320	No specimens identified	407 / 519	No specimens identified	No specimens identified	F / 245 (330)	F / 312 (353)	F / 284 / 314
		37 / C	35 / A	20 / C/Ax3		34 / A			18 (20) / Ax2 (C/Ax1)	22 (24) / Ax2 (C/Ax1)	21 / Ax3 / 22 / C/Ax3
PU6/Ep		No specimens received	F / 152	F / 108 (121)	263 / 251	270 / 191	F / 140	F / 163	F(6) / 99	F / 116	236 / 208 (293)
			18 / C/Ax12	13 (14) / C/Ax4 (Ax2)	25 / A	20 / Ax6	15 / C/Ax9	17 / C/Ax9	11 / C/Ax6	11 / C/Ax6	24 (25) / C/Ax10 (Cx2)
PSI/Ep		90 / 349 (326)	67 / 137 (156)	68 / 222	63 / 204	67 / 239	No specimens identified	72 / 197	63 / 137	67 / 188	58 / 135
		579 (536) / Cx2 (C/Ax1)	172 (270) / Ax2 (C/Ax1)	486 / C/A	421 / C/A	475 / C/A	F / 20	395 / C/A	215 / C/A	358 / C	223 / C/Ax6
PSI/Isocy		75 / 197 (280)	F / 19	62 / 125	64 / 116	62 / 83	8 / A	54 / 63	F / F	60 / 94	F / 32
		217 (468) / Ax2 (C/Ax1)	7 / A	143 (87–214) / A	107 / Ax2	58 / A		36 / A	F / A	75 / A	18 / Ax6
PS3/Ep		No specimens received	76 / 101	87 / 113	260 / 240 (281)	82 / 123	64 / 106 (61–155)	60 / 83	12 / 13	54 / 65	46 / 58
			53 / Ax12	56 / Ax5	25 (26) / Ax2 (C/Ax1)	56 / Ax6	89 (22–150) / A	65 / Ax6	39 / C/Ax6	41 / C/Ax6	43 / C/Ax12
Si2/Acryl		243 / 504	218 / 331	212 / 323 (338)	207 / 359	197 / 331	182 / 301	210 / 389	183 / 345	220 / 343	197 / 425
		129 / C	98 / C	53 (87) / C/Ax2 (Cx1)	106 / C	110 / C	106 / C	128 / C	119 / C	61 / C	135 / Cx6

Key

F_{25}	F_{max}		u	unacceptable		A	failure at the sealant/primer interface		x	number of samples showing a particular mode of failure
E_{max}	Failure mode		0	least acceptable		C	failure in the bulk of the sealant			
			3	most acceptable		C/A	failure close to the interface			

CIRIA C520

Table A5.3 *Change in properties compared to dry controls*

		Dry	Wet	Soil	Yorks pre	Yorks post	Tewkes pre	Tewkes post	99% humidity	Sewage outfall	Sewage anaerobic
PU3/Ep		423	-3%	-5%	-41%	+10%	-20%	-5%	-23%	-17%	N/R
		493	-19%	-23%	-13%	-26%	-20%	-22%	–	-29%	-28%
		C	C/A	C	C	C	C	C	A	A	C
		43	+25%	+30%	+7%	+16%	-9%	+2%	-19%	-16%	N/R
		45	0%	-22%	-13%	+9%	-11%	-18%	-51%	-53%	-16%
		C	C	A	C	C	C	C	A	A	C
PU3/Isocy		427	-13%	-19%	-17%	-10%	-21%	-11%	-24%	-24%	N/R
		417	-9%	–	–	-2%	N/R	N/R	–	–	–
		C	C	A	C	C	N/R	N/R	A	A	C
		43	+19%	-16%	-37%	-37%	-37%	-35%	-44%	-40%	-43%
		37	-5%	-46%	N/R	-8%	N/R	N/R	-51%	-41%	A
		C	A	C/A	N/R	A	N/R	N/R	A	A	A
PU6/Ep		N/R	N/R	–	–	–	–	–	–	–	–
		N/R	N/R	–	–	–	–	–	–	–	–
		N/R	N/R	–	–	–	–	–	–	–	–
		N/R	N/R	–	–	–	–	–	–	–	–
		N/R	N/R	–	–	–	–	–	–	–	–
		N/R	N/R	–	–	–	–	–	–	–	–
PSl/Ep		83	-18%	-24%	-43%	-16%	-13%	-24%	-31%	-28%	C/A
		90	-26%	-24%	-30%	-26%	N/R	-20%	-30%	-26%	-36%
		C/A	C/A	A	A	A	C/A	C/A	A	A	A
		528	-22%	+7%	-14%	-1%	-30%	-26%	-30%	-32%	-62%
		579	-70%	-16%	-27%	-18%	N/R	-32%	-63%	-38%	C/A
		C	A	C	A/C	C/A	N/R	C/A	A	C	C/A
PSl/Isocy		78	–	-17%	-27%	-14%	–	-27%	–	-23%	–
		75	–	-34%	-15%	-17%	–	-28%	–	-20%	–
		C/A	–	A	A	A	–	A	–	A	–
		522	-96%	-61%	-91%	-88%	-96%	-92%	-98%	-84%	-92%
		217	-97%	–	-51%	-73%	-96%	-83%	–	-65%	A
		A	A	A	A	A	A	A	A	A	A
PS3/Ep		N/R	N/R	–	–	–	–	–	–	–	–
		N/R	N/R	–	–	–	–	–	–	–	–
		N/R	N/R	–	–	–	–	–	–	–	–
		N/R	N/R	–	–	–	–	–	–	–	–
		N/R	N/R	–	–	–	–	–	–	–	–
		N/R	N/R	–	–	–	–	–	–	–	–
Si2/Acryl		223	-10%	-9%	-10%	-15%	-19%	-5%	+18%	-7%	C/A
		243	–	-13%	-18%	–	-25%	-14%	-25%	-10%	-19%
		C	C	A	C	A	A	A	A	A	C/A
		156	+24%	+14%	+37%	-2%	-8%	+5%	+10%	+12%	-5%
		129	-24%	-59%	-18%	-15%	-18%	-1%	-8%	-53%	C
		C	C	C/A	C	C	C	C	C	C	C

Key

TP1	TP2			
ΔF_{25}	ΔF_{25}	u	unacceptable	A failure at the sealant/primer interface
ΔE_{max}	ΔE_{max}	0	least acceptable	C failure in the bulk of the sealant
F mode	F mode	3	most acceptable	C/A failure close to the interface

Note Absolute values are given for the dry controls

Table A5.4 *Change in properties compared to wet controls*

		Dry	Wet	Soil	Yorks pre	Yorks post	Tewkes pre	Tewkes post	99% humidity	Sewage outfall	Sewage anaerobic
PU3/Ep	TP1 ΔF_{25}	+3%	410	-2%	-39%	+13%	-18%	-2%	-21%	-15%	-13%
	TP1 ΔE_{max}	-20%	54	+4%	-15%	-7%	-28%	-19%	-35%	-33%	-53%
	TP1 F mode	C	C/A	C	C	C	A	C	A	A	A
	TP2 ΔF_{25}	+15%	373	-7%	-5%	+3%	-10%	-2%	-13%	-13%	-37%
	TP2 ΔE_{max}	-16%	51	-29%	-47%	-47%	-47%	-45%	-53%	-49%	—
	TP2 F mode	C	A	A	C/A	A	A	A	A	A	A
PU3/Isocy	TP1 ΔF_{25}	N/R	N/R	—	—	—	—	—	—	—	—
	TP1 ΔE_{max}	N/R	N/R	—	—	—	—	—	—	—	—
	TP1 F mode	N/R	N/R	—	—	—	—	—	—	—	—
	TP2 ΔF_{25}	N/R	18	-28%	+39%	+11%	-17%	-6%	-39%	-39%	+33%
	TP2 ΔE_{max}	N/R	C/A	C/A	A	A	C/A	C/A	A	C/A	C/A
	TP2 F mode										
PU6/Ep	TP1 ΔF_{25}	+22%	68	+2%	-31%	+3%	+6%	-7%	-16%	-12%	-13%
	TP1 ΔE_{max}	+34%	67	-7%	-6%	0%	-10%	+8%	-6%	-12%	+108%
	TP1 F mode	C/A	A	A	A	A	C/A	C/A	A	A	0%
	TP2 ΔF_{25}	+29%	411	+38%	+10%	+27%	-10%	-4%	-10%	+25%	+30%
PSI/Ep	TP1 ΔF_{25}	+237%	172	+183%	+145%	+176%	-145%	+130%	—	+108%	C/A
	TP1 ΔE_{max}	C/A	A	C	C/A	C/A	C/A	C/A	C/A	C	—
	TP1 F mode	—	—	—	A C/A	—	—	—	—	—	—
PSI/Isocy	TP1 ΔF_{25}	+2400%	21	+870%	+114%	+200%	-10%	+110%	48%	+295%	—
	TP1 ΔE_{max}	+3000%	7	+1943%	+1430%	+729%	+14%	+414%	—	+971%	+157%
	TP1 F mode	A	A	A	C/A	A	A	A	A	A	N/R
PS3/Ep	TP1 ΔF_{25}	N/R	N/R	+15%	+242%	+8%	-16%	-21%	-84%	-29%	-39%
	TP1 ΔE_{max}	N/R	76	+6%	-53%	+6%	+68%	+23%	-26%	-23%	-19%
	TP1 F mode	N/R	53	A	A	—	A	—	—	C/A	C/A
	TP2 ΔF_{25}	N/R	A		C/A						
Si2/Acryl	TP1 ΔF_{25}	-19%	200	+2%	0%	-5%	-10%	+7%	+32%	+4%	-10%
	TP1 ΔE_{max}	+12%	193	-8%	+11%	-21%	-26%	-16%	-11%	-10%	+38%
	TP1 F mode	+32%	98	-46%	+8%	+12%	+8%	-4%	+21%	-38%	C
	TP2 ΔF_{25}	C	C	C/A	C	C	C	+31%	C	+1%	C

Key

TP1	TP2			
ΔF_{25}	ΔF_{25}	u	unacceptable	A failure at the sealant/primer interface
ΔE_{max}	ΔE_{max}	0	least acceptable	C failure in the bulk of the sealant
F mode	F mode	3	most acceptable	C/A failure close to the interface

Note Absolute values are given for the wet controls

Table A5.5 WIS rating against dry controls

Sealant		Dry	Wet	Soil	Yorks pre	Yorks post	Tewkes pre	Tewkes post	99% humidity	Sewage outfall	Sewage anaerobic
PU3/Ep	TP1 rating		1	1	u	2	u	1	u	0	u
	TP1 F mode	C	C/A	C	C	C	A	C	A	A	N/R
	TP2 rating		1/0	u	0	0	u	u	u	u	u
	TP2 F mode	C	C	A	C	C	A	C	C/A	A	C
PU3/Isocy	TP1 rating		1	0	–	u	u	0/u	u	u	u
	TP1 F mode	C	A	A	N/R	A	A	A	A	A	N/R
	TP2 rating		0	u	N/R	u	N/R	N/R	u	u	u
	TP2 F mode	C	A	C/A	N/R	A	N/R	N/R	A	A	A
PU6/Ep	TP1 rating		–	–	–	–	–	–	–	–	–
	TP1 F mode	N/R	N/R	A	–	–	–	–	–	–	–
	TP2 rating		–	C/A	–	–	–	–	–	–	–
	TP2 F mode	N/R	C/A	A	A	A	C/A	C/A	C/A	C/A	C/A
PSI/Ep	TP1 rating		0	0	u	0	0	u	u	u	u
	TP1 F mode	C/A	A	C	A/C/A	C/A	C/A	C/A	A	C	C/A
	TP2 rating		u	u	u	u	u	u	u	u	u
	TP2 F mode	C	A	C/A	C/A	C/A	C/A	C/A	C/A	C	C/A
PSI/Isocy	TP1 rating		u	u	u	u	u	u	u	u	u
	TP1 F mode	C	A	A	C/A	A	A	A	A	A	A
	TP2 rating		u	0	u	u	u	u	u	u	u
	TP2 F mode	A	A	A	A	A	A	A	A	A	A
PS3/Ep	TP1 rating		–	–	–	–	–	–	–	–	–
	TP1 F mode	N/R	N/R	A	C/A	A	A	A	A	C/A	C/A
	TP2 rating		–	–	–	–	–	–	–	–	–
	TP2 F mode	N/R	N/R	A	C/A	A	A	A	C/A	C/A	C/A
Si2/Acryl	TP1 rating		1	1	u	u	0	0	2	1	0
	TP1 F mode	C	C	C	C	C	C	C	C	C	C
	TP2 rating		0	u	u	0	u	u	u	u	u
	TP2 F mode	C	C	C/A	C	C	C	C	C	C	C

Key

TP1	TP2		u unacceptable	A failure at the sealant/primer interface
WIS rating	WIS rating		0 least acceptable	C failure in the bulk of the sealant
F mode	F mode		3 most acceptable	C/A failure close to the interface

Table A5.6 WIS rating against wet controls

	Dry	wet	Soil	Yorks pre	Yorks post	Tewkes pre	Tewkes post	99% humidity	Sewage outfall	Sewage anaerobic
PU3/Ep	1	1	1	u	1	0	0	u	0	0
	C	C/A	C	C	C	A	C	A	A	N/R
	1/2	C	0	1	1	0	0	u	u	0
	C	C	A	C	C	C	C	C/A	A	C
PU3/Isocy	1		u	u	u	u	N/R	u	u	u
	C	A	A	C/A	A	A	N/R	A	A	N/R
	2	A	0	N/R	1	N/R	N/R	u	u	u
	C		A	N/R	A	N/R	N/R	A	A	N/R
PU6/Ep	N/R	N/R	–	–	–	–	–	–	–	–
	N/R	C/A	A	A	A	C/A	C/A	A	A	N/R
	N/R		–	–	–	–	–	–	u	u
	N/R	A	C/A	A	A	C/A	C/A	C/A	C	A
PSI/Ep	2		3+	0	2	1	3+	0	0	1
	C/A	A	C/A	A C/A	C/A	C/A	C/A	A	A	N/R
	3+	A	2	2+	3+	1	3+	1	3+	1
	C		A	C/A	C/A	C/A	C/A	C/A	C	C/A
PSI/Isocy	3+	A	(3+)	(3+)	(3+)	–	(3+)	u	(3+)	(3+)
	C	A	A	A	A	A	A	A	A	A
	3+	A	(3+)	(3+)	(3+)	–	(3+)	–	(3+)	(3+)
	A		A	C/A	A	A	A	A	A	A
PS3/Ep	N/R	N/R	–	–	–	–	–	–	–	u
	N/R	A	A	C/A	A	A	A	A	C/A	N/R
	N/R		2	u	2	2	0	u	u	u
	N/R	A	A	A	A	A	A	A	A	N/R
Si2/Acryl	1	1	1	2/1	0	0	1	1+	1	2
	C	C	C	C	C	C	C	C	C	N/R
	2		u	1	1	1	1	1	u	2
	C	C	C/A	C	C	C	C	C	C	C

Key

TP1
WIS rating
F mode

TP2
WIS rating
F mode

u unacceptable
0 least acceptable
3 most acceptable

A failure at the sealant/primer interface
C failure in the bulk of the sealant
C/A failure close to the interface